DIRECTIONS IN SAFETY-CRITICAL SYSTEMS

Proceedings of the First Safety-critical
Systems Symposium

The Watershed Media Centre, Bristol
9–11 February 1993

W0055404

Edited by
FELIX REDMILL and TOM ANDERSON

Springer-Verlag
London   Berlin   Heidelberg   New York
Paris   Tokyo   Hong Kong
Barcelona   Budapest

Felix Redmill
Redmill Consultancy
22 Onslow Gardens
London N10 3JU, UK

Tom Anderson
Centre for Software Reliability
University of Newcastle-upon-Tyne
20 Windsor Terrace
Newcastle-upon-Tyne NE1 7RU, UK

ISBN-13:978-3-540-19817-8   e-ISBN-13:978-1-4471-2037-7
DOI: 10.1007/978-1-4471-2037-7

British Library Cataloguing in Publication Data
A catalogue record for this book is available from the British Library

Library of Congress Cataloging-in-Publication Data
A catalog record for this book is available from the Library of Congress

Typesetting: Camera ready by author

34/3830-543210 Printed on acid-free paper

PREFACE

The contributions to this book were all invited in order to form a structured and balanced programme for the Safety-Critical Systems Club's first annual symposium. The symposium itself was planned so as to cover a wide range of issues which, in turn, were grouped under three themes, one for each day of the event.

The first day's theme, *Experience from Around Europe*, reminds us that when it comes to safety – indeed, more generally, when it comes to technology – we are not alone, not isolated. We protect others by our safety measures and are protected by theirs. The technologies that each develops, if shared, will enable us all to do a better job of protecting the earth. The contributions from Germany, The Netherlands, and Norway report on the experience of many years in the achievement and assessment of safety; the final contribution is an early paper from a new quarter – the sociological viewpoint. There will, I think, be an increasing interest in the safety domain by social scientists, for safety depends not only on technology, but also on human and sociological issues. The Human Factors experts are already at work, now here come the Sociologists. While the wiser of them may feel the need for a period of acclimatization before arriving at certainty, we could usefully heed their ideas and ways of looking at things.

The second day's theme, *Current Research*, encompasses eight large research projects in the UK. The contributions are not dull or esoteric reports written in language accessible only to academics, for all the projects are collaborative between academia and industry, and all are intended to produce technologies immediately applicable to industrial problems or improvements. The contributions set out the issues which the projects are designed to address and the safety-enhancing techniques and methods which are being sought. They will therefore inform academics of research which is complimentary to their own and industrialists of the technologies which will soon be available and where to find them. These technologies are expected to be widely applicable, for there is cross-sector collaboration in almost all the projects.

On the third day, *Achieving and Evaluating Safety* is explored. The first paper considers safety arguments, a subject of increasing importance. Suppose the presentation of the safety cases were standardized? Wouldn't this ease the burden both on those who prepare them and those who adjudicate on them? And wouldn't it, in doing so, enhance safety? Can we afford not to move towards standards in this area? Other contributions explore how methods already in use in other domains may be applied to safety, and examine the data requirements for a convincing safety evaluation. A broad spectrum is covered in this part of the book, and the contributions will be informative not only to experts in the safety domain but also to novices as an introduction to the breadth of the field.

My acknowledgement and gratitude go to all the authors whose contributions not only made this book possible but also comprised an important symposium. The book's publication was achieved only through their willingness to fit the preparation of their contributions into already encumbered schedules. Their cooperation during the preparation and editing phases is greatly appreciated.

Finally, I would like to express my thanks to Bob Malcolm for his considerable help and advice in planning the programme.

Felix Redmill
November 1992

CONTENTS

The Safety-Critical Systems Club
sponsor and organiser
of the
Safety-critical Systems Symposium

The Safety-Critical Systems Club exists for the benefit of industrialists and researchers involved with computer systems in safety-critical applications. Sponsored by the Department of Trade and Industry and the Science and Engineering Research Council, its purpose is to increase awareness and facilitate communication in the safety-critical systems domain.

The British Computer Society and the Institution of Electrical Engineers are active participants in the management of the Club, the Centre for Software Reliability at the University of Newcastle upon Tyne is responsible for organisation, and Felix Redmill, an independent consultant, is the Club's Co-ordinator.

The Club's goals are to achieve, in the supply of safety-critical computer systems:

- Better informed application of safety engineering practices;
- Improved choice and application of computer-system technology;
- More appropriate research and better informed evaluation of research.

In pursuit of these goals, the Club seeks to involve both engineers and managers from all sectors of the safety-critical community and, thus, to facilitate the transfer of information, technology, and current and emerging practices and standards.

The club publishes a newsletter three times per year and organises and co-sponsors events on safety-critical topics. In these ways it facilitates: the communication of experience among industrial users of technology, the exchange of state-of-the-art knowledge among researchers, the transfer of technology from researchers to users, and feedback from users to researchers. It facilitates the union of industry and academia for collaborative projects, and it provides the means for the publicity of those projects and for the reporting of their results.

To join or to enquire about the Club, please contact:
Mrs J Atkinson
Centre for Software Reliability
The University
20 Windsor Terrace
Newcastle upon Tyne, NE1 7RU
Tel: 091 212 4040; Fax: 091 222 7995.

PART I

EXPERIENCE FROM AROUND EUROPE

Tuesday 9th February 1993

Certification of Safety Critical Systems in Germany

Gerhard Rabe

Technischer Überwachungsverein Norddeutschland e.V.
Hamburg, Germany

Abstract

The Technical Inspection Associations "TÜV" in Germany can look back on more than 100 years of operation as inspection and consultancy organisations. There are now twelve TÜV distributed around the country that work independently of one another. They are totally committed to impartiality and are independent of pressure groups. They operate on a non-profit basis.

Their activities cover a wide range of technical fields encompassing safety, energy, traffic and ecology. The structure of the TÜVs will be presented in detail.

In Germany, safety critical applications must be certified by independent assessors before they are licensed by the authorities. The certification, which is conducted according to state of the art principles, is performed almost exclusively by the TÜV's.

The procedures for the certification of safety critical systems will be described and basic rules presented.

1 Introduction

In Germany, as in many other countries, technical equipment and plants that can represent a hazard for people or the environment must be licensed in accordance with legal prescriptions. Licenses are granted by the appropriate authorities based on the applicable laws and standards. These authorities are normally state (rather than federal) authorities in Germany.

In the last few years use of computers in safety relevant applications has increased considerably. Correspondingly, such systems must be taken into consideration in the licensing procedure. Examples are:

- steering rod control computers in nuclear power plants,
- trouble logging computers in nuclear power plants,
- pipeline control computers in the petrol industry,
- computers for controlling PVC production in the chemical industry,
- control computers for respirators.

In the licensing procedure, the authorities employ consultancy organisations for the assessment of safety relevant systems. One of these consultancy organisations is the Technical Inspection Association of Northern Germany, TÜV Norddeutschland (TÜV-N), which along with the other eleven TÜVs is one of the best known assessment organisations.

The TÜV's tasks are not restricted to the assessment of computer systems: the activities encompass civil engineering, energy technology, environmental protection and automobile technology.

In the following, the history of the TÜVs and their structure will be described. The individual fields of activity will be described in greater detail. The certification process and the involvement of the TÜV will be explained. This will be done based on two examples from computer technology.

Existing laws and standards form the basis for the licensing procedure. At the end of the article, the DIN 19250 [DIN 89a] will be examined in more detail, because it can be regarded a universal basis for the use of measuring and control devices.

2 The nature of TÜV

The origins of the TÜVs lie in the days when steam engines were being increasingly deployed in industrial plants, i.e. late in the previous century. The new technology brought with it considerable risks: the yet inadequate plants presented considerable hazards and accidents caused considerable damage.

When the number of accidents - particularly with boilers - reached a level that could no longer be accepted, the industry gave in to pressure from the authorities (trade regulations) and decided to build up its own inspection organisation. The goal was to have the plants regularly checked by experts and thus increase their reliability and safety.

This was the birth of the TÜVs, those days known as "Boiler Inspection Associations". There arose a number of inspection organisations with regional responsibility districts. This meant that there was no central organisation so that the associations began to compete with one another as deregulation proceded.

Today there are twelve TÜVs in Germany with the status of registered associations. The associations work as non-profit organisations and are independent of pressure groups. The members of the associations are industrial companies represented in the association's presidium and thus in a position to influence the global industrial policy. They cannot influence the professional assessments.

The basis for the TÜVs' activities was originally the Trade Ordinance [GewO 90], whose §24 prescribes the inspection of boilers. The collection of standards has been developed continuously in the course of the years and today the Atomic Act and the Road Traffic Authorisation Ordinance for automobiles form along with the Trade Ordinance the basis for our work.

In short, the purpose and goal of the Technical Inspection Associations can be described as follows:

- the technical examination of plants, equipment and resources as well as training and examination of persons, in so far as these activities are required by official rules and responsibility has been transferred to the TÜV;
- performance of consultancy tasks as well as appropriate training and examination tasks for individuals;
- collection of technical, civil and energy economical experience.

3 Fields of work of the TÜV

A hundred years ago, protection of people and assets against the risks and inadequacies of steam engines was of primary importance. Soon advice on the economical deployment of resources in energy production plants and ultimately other technical equipment came in addition to inspection. At the present time there is also an attempt to not only improve the safety and economy of the plants, but also to improve their reliability and to begin humanising technology.

The individual fields of work can be recognised using the organisational structure of TÜV-N as an example. The association is subdivided into three principle sectors:
- industrial plant technology,
- energy technology, material technology and environmental protection,
- automobiles.

Outside the automobile sector there is a further subdivision into the departments:
- Steam and Pressure,
- Electrical and Transport Technology,
- Heating and Tanks,
- Material Technology,
- Nuclear Power and Radiation Protection,

- Environmental Protection.

In all these fields the TÜV-N is not only active as an assessor, but also provides consultancy for companies and authorities. Thanks to its universal capability of providing assessment and consultancy, the TÜV helps in adhering to high safety standards. This means that corresponding products can not only be licenced in Germany, but in the whole of Europe.

This spreads the activities wide outside the association's regional boundaries. Our house has dependencies in Scandinavia, the activities in the east of Europe are being increased. The tasks are also performed in credited test laboratories (in accordance with the EN 45000 series [DIN 90a]). For the production processes an internal quality assurance system according to ISO 9000 [ISO 90] has been established. The TÜV-N also works on national and international regulatory committees.

The high safety standard that had been developed in the past has been transferred to today's technologies; particularly in the field of nuclear technology the standard is regarded as exemplary. This has been substantially achieved by the TÜVs.

4 Certification procedures

The basic principles of the the certification process will be demonstrated with two examples from the nuclear field and one from a non nuclear field.

4.1 The nuclear licensing procedure

Nuclear regulations (amongst others [KTA 85]) prescribe that systems with safety tasks shall be qualified before they are employed in a nuclear power plant. When a plant operator wants to use a newly

7

developed system, the manufacturer of the system contacts an assessment institution which in most cases is a TÜV. In negotiations the existing regulations are implemented in such a way that the requirements for the qualification and the extent of testing can be defined. This is finally set down in a contract.

If the newly developed system is a computer system, both the hardware and the software are subjected to the qualification process. Within the scope of this paper the details of the individual test steps will not be described; for more detail, see [Rabe 89], [Anders 90] and [Glöe 90].

The test process can be summarised as follows:
- System analysis;
- Hardware examination (theoretical, practical);
- Software examination (theoretical, practical with special tools);
- Integration test.

When the tests have been passed, the test procedure and results are recorded in a test report. For the tested system a certificate is awarded that allows the manufacturer to offer the system for safety critical applications.

It is important to note here that this type of testing is a plant independent type approval that is accepted by all the licensing authorities in Germany.

This is the first step towards certification. The operator of a nuclear power plant must apply to the licensing authorities for a licence. Since he can present a certificate for the type approval, only the plant specific adequacy needs to be demonstrated. To this end the licensing authorities normally employ the TÜV.

The examination that now takes place is mainly based on the prescriptions resulting from the processual or hazard analysis and concentrates on the suitability of the device for the specific application. This can also involve examination of documenation (e.g. value tables for parameters). At any event, there is an on the spot test covering not only the physical installation (e.g. cabeling) but also - and that is most important - the correct functioning of the system.

Also this test is documented in a test report that is presented to the licensing authority. Based on the conclusions stated in the test report, the licensing authority makes the final decision on employment of the corresponding system (for which it is responsible).

A license can be granted with a limited duration and can contain additional prescriptions such as
- prescriptions for regular reexaminations
- prescriptions to monitor and evaluate the operational behaviour of particular parts.

4.2 The non nuclear licensing procedure

In the non nuclear field the certification process differs slightly from the procedure described above. It is important to point out here that one can differentiate between
- a certification on behalf of the authorities of a plant subject to inspection,
- a certification as a general quality evidence for any application in order to open up a market for the product.

In most cases systems in safety critical applications are tested after their installation in the plant. The examination can then be subdivided into a purely system related examination, covering hardware and software, and a plant specific examination. As in the nuclear licensing procedure, the license is granted by the appropriate authorities based on the recommendations of test reports from the TÜVs.

In Germany there are at present only a few standards that can be applied directly to the certification of safety critical plants. Mention should be made of the DIN VDE 0116 [DIN 89b], which also makes very concrete statements on the examination. The central point of the standard states that the system shall be safe against primary failures. All further requirements and tests are derived from this.

The IEC standards [IEC 91] and [IEC 92] are also taken into consideration.

As a universal basis for the use of measurement and control systems one can take DIN 19250, which will be described in more detail in the following. DIN VDE 801 [DIN 90b] should also be mentioned here.

5 Basic rules

The DIN 19250 is a pre standard intended for standardisation committees that are concerned with defining safety technical rules for the reliable function of measuring and controlling protection systems.

Implementation of this standard proceeds in three steps:
- estimation of the hazard to be covered;
- definition of requirements;
- allocation of technical and organisational measures.

For the estimation of the hazard that shall be covered a qualitative process is described that leads to graduated requirements classes. For these requirements classes, bundles of technical and organisational measures for the equipment of the measurement and control system are determined that will ensure that the risk will at most reach the defined limits or preferably remain below them.

5.1 Risk estimation

In order to estimate the risk that a safety critical plant represents, one starts by examining independent individual risks. Via the individual risks and individual protection measures one can determine the risk of the total system.

An exact quantification of the risk is often not possible, or at least it involves considerable effort and is extremely difficult. For simplification, DIN 19250 introduces four parameters to describe the risk in the event of failure or non existence of the measuring and controlling protection system.

According to DIN VDE 31000 part 2 [DIN 87], the risk R is defined as the combination of factors

$$R = F \times H$$

where F is the Frequency of harmful events and H is the Harm they cause. The factor F results from:
- the Duration D of presence in the danger zone
- the Ability A to avert danger
- the Probability P of the dangerous event without the presence of the measuring and controlling protection system.

These four factors can be graded as follows:

- Duration in the danger zone (D)
    D1 seldom to often
    D2 frequently to permanently
- Aversion of danger (A)
    A1 possible under certain conditions
    A2 virtually impossible
- Probability of the dangerous event (P)
    P1 very small
    P2 small
    P3 relatively large

11

- Harm caused by the event (H)
  H1 light injuries
  H2 heavy, irreversible injury of one or more persons or death of a single person
  H3 several deaths
  H4 catastrophe, extremely high death toll

Based on this there are theoretically a total of 48 possible graduations. However, due to the dominance of certain risk parameters, only eight are of significance. For extremely harmful consequences (Harm grade H4) it turns out that the parameters D and in particular A play a negligible role.

The possible combinations can be represented more clearly in a risk graph:

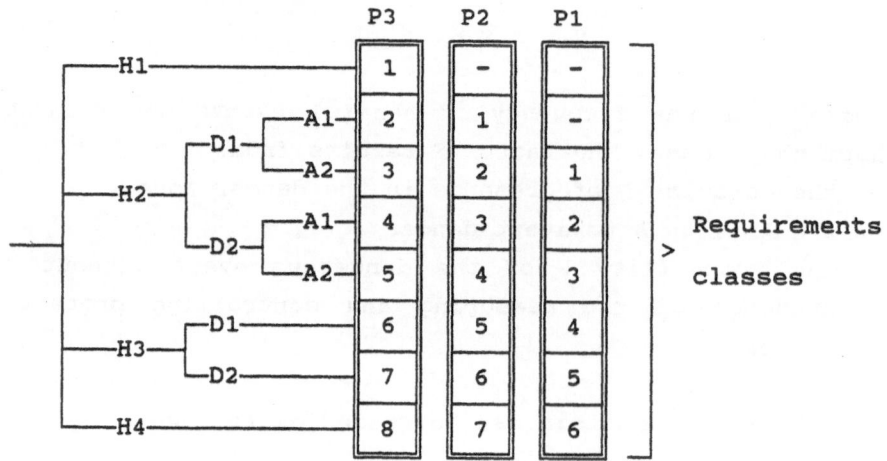

With the aid of this graph a measure of the risk to be covered can be derived from largely objective criteria. This measure is divided into eight grades, the so called requirements classes. The higher the requirements class is, the higher the partial risk to be covered is and the higher the requirements.

5.2 Definition of requirements

In DIN 19250 the requirements classes are associated with various possible requirements. These range from error handling to demands on operational and environmental conditions.

The following table gives an extract of an exemplary allocation of properties to requirements classes:

Examples of properties for which requirements can be defined	Requirements class							
	1	2	3	4	5	6	7	8
1a Random errors	(X)	(X)	(X)	X	X	X	X	X
1b Random errors (rare)							X	X
2a Systematic errors, avoidable					(X)	(X)	X	X
. . .								. . .
7 Repair/Maintenance/ Modification ° normal ° restrictive	(X)	(X)	(X)	(X)	(X) (X)	X	X	X
8 Operational and en- vironment conditions ° within scope ° outside scope	X	X	X	X	X	X	X (X)	X X
X is normally required (X) is only partially considered resp. required								

5.3 Allocation of technical and organisational measures.

The bundles of measures for measurement and controlling protection systems, graded by requirements class, consist of technical and organisational measures. The various kinds of measure can supplement or replace one another. This means that the safety of the system can be achieved in a number of equivalent ways.

As examples of measures the standard mentions amongst others:
- self monitoring
- redundancy
- regular inspections.

In summary it can be said that, on the basis of this standard, the systems to be certified can be classified and the relevant requirements determined. The test criteria and procedures are defined based on these requirements.

6 Summary

The experience of more than a hundred years has shown that the work of the TÜVs in the field of safety technology is sensible and necessary. In this way it can be guaranteed that a high standard of safety and reliability can be achieved in order to avert harm from people, the environment and industrial plants.

The TÜVs have shown themselves to be reliable, neutral and expertly qualified partners. The certification of safety critical systems that they perform support both the manufacturers and operators as well as the licensing authorities.

The procedures and principles that are employed in the certification process have shown themselves to be most effective.

The TÜVs orientate themselves towards the expansion of the european market as envisaged by the European Treaty. They see their task in providing their capabilities and experience in and outside Europe. They regard this as a contribution to overcoming the still present hurdles that hinder the exchange of goods and services.

References

[Anders 90] Anders U, Mainka E U, Rabe G: "Tools and methodologies for quality assurance", Safecomp'90, Gatwick, November 1990

[DIN 87] DIN VDE 31000, Teil 2: "Allgemeine Leitsätze für das sicherheitsgerechte Gestalten technischer Erzeugnisse ; Begriffe der Sicherheitstechnik; Grundbegriffe (General principles for the safe design of technical products; basic concepts)", December 1987

[DIN 89a] DIN 19250: "Grundlegende Sicherheitsbetrachtungen für MSR-Schutzeinrichtungen (Fundamental safety aspects to be considered for measurement and control equipment)", January 1989

[DIN 89b] DIN VDE 0116: "Elektrische Ausrüstung von Feuerungsanlagen (Electrical equipment in firing stations)", October 1989

[DIN 90a] DIN EN 45001: "Allgemeine Kriterien zum Betreiben von Prüflaboratorien (General criteria for the operation of testing laboratories)", May 1990

[DIN 90b] DIN V VDE 801: "Grundsätze für Rechner in Systemen mit Sicherheitsaufgaben (Principles for computers in safety-related systems)", January 1990

[GewO 90] Gewerbeordnung, November 1990

[Glöe 90] Glöe G, Rabe G: "Tools for quality assurance for controller software", ISATA Symposium on Automotive Technology and Automation, Florence, May 1990

[IEC 91] IEC 65A WG9: "Software for computers in the application of industrial safety-related systems", November 1991

[IEC 92]   IEC 65A WG10: "Draft: Functional safety of electrical/electronic/programmable electronic systems: General aspects", 1992

[ISO 90]   ISO 9000: "Quality management and quality assurance standards", May 1990

[KTA 85]   Kerntechnischer Ausschuß: "KTA 3501 - Reaktorschutzsystem und Überwachungseinrichtungen des Sicherheitssystems (Reactor protection system and control equipment of the safety system)", June 1985

[Rabe 89]   Rabe G: "Safety requirements in process automation", VTT Symposium on safety in the design of processes and production automation, Tampere, May 1989

SAFETY FOR EUROPEAN SPACE AGENCY SPACE PROGRAMMES.

Keith M. Wright
Head of Safety Section,
European Space Agency,
European Space Technology Centre,
Noordwijk, The Netherlands.

Abstract.

Safety for ESA Space programmes is implemented as part of the ESA Product Assurance and Safety discipline. An ESA Safety Policy, Safety Programme, and supporting implementation and technical system safety requirements have been established. The ESA Safety programme comprises the systematic identification and evaluation of the hazardous characteristics and risks associated with space systems design, operation, and operating environment, together with a process of safety optimisation through hazard and risk reduction and management. The space system which is addressed is considered to comprise hardware, software, and the system operator.

1. Introduction.

In 1985 the member states of the European Space Agency approved a new Long Term Space Programme. Included in this Programme is the development of autonomous facilities for the support of man in space, for the transport of equipment and crews, and for making use of low Earth orbit. Although the content of the programme has since been modified, its principle objects remain the same.

At the time of the approval of the Long Term Space Programme, ESA's safety requirements were aimed at supporting the development of unmanned scientific and applications technology satellites, and as such were considered to be inadequate for the new Long Term Space Programme. However ESA's Spacelab programme, performed in support of NASA's development of the NASA Space Transportation System - Space Shuttle (NSTS), and also ESA's involvement in the development of some payloads for the NSTS, had provided some direct experience with manned spaceflight safety programmes. Based on this experience, systematic research was commenced with the objective of defining technically effective and cost effective Product Assurance and Safety Requirements which would be able to support the objectives of the Long Term Space Programme.

The research comprised the evaluation of past experience in Europe and the U.S.A. with the Safety of manned spaceflight, and the evaluation of the safety policy and requirements which have been established and implemented for selected "high risk" non-space industries. These industries were: off-shore oil; chemical processing; nuclear power; civil aviation; and sea transportation. From this research, and from ESA's own experience, a Product Assurance and Safety policy has been developed and requirements have been established, which are intended to effectively support the initial phase of the Long Term Space Programme.

The ESA Safety Policy and Requirements are documented in the PSS-01 series of the ESA's Procedures, Standards, and Specifications documentation system. The ESA PSS-01 series of documents address the Product Assurance and Safety disciplines and are structured in three levels:

- Level I,   Policy;

- Level II,   Requirements;

- Level III, Supporting Requirements and Implementation Methods.

The total set of Product Assurance and Safety documents addresses: Quality Assurance; Configuration Control; Reliability; System Safety; Maintainability; Software Assurance; and Parts, Materials and Processes requirements and application control. Figure 1 shows Levels I and II of the PSS-01 documentation tree, while Figure 2 shows Levels II and III of the Safety documentation tree.
   A Product Assurance Database and computerised safety analysis support tools are also currently being developed in order to support the effective implementation of these requirements.

2. Responsibility.

Responsibility for assuring the implementation of System Safety in ESA is vested in the Product Assurance and Safety Department of the ESA Technical Directorate. The Department is responsible for the development and maintenance of the Product Assurance and Safety standards (the PSS-01 documents) and for providing functional support to ESA programmes and projects in its areas of expertise. The Department organisation is shown in Figure 3.
   In order to maintain its technical competence and to provide a basis for the development or upgrading of requirements, procedures, and tools, the Department manages several programmes of research, and maintains Materials and Processes, and Electrical and Electronics Parts technology laboratories.
   The Product Assurance policy and requirements which are documented in the PSS-01 series of specifications are applicable to all ESA programmes and projects. These requirements are "tailored" to the specific needs of each, and formally baselined.
   Project dedicated Product Assurance and Safety teams are established as necessary, and are integrated into the ESA Project team. They are functionally responsible to the Project Manager for day to day Project activities, and are responsible to the Head of the Product Assurance and Safety Department for Policy matters. It is their responsibility to oversee the application and implementation, by the Project Contractors, of the tailored Product Assurance and Safety Requirements. Project milestone reviews are supported through the participation of members of the Product Assurance and Safety Department. A degree of independent assurance authority is thereby maintained.

3. ESA Safety Policy.

Spaceflight activities, and in particular manned spaceflight activities are generally acknowledged to be inherently risky. As a public organisation supported by public funds, ESA and its programmes, as well as its astronauts, can be threatened by the risks to which the Agency is exposed in the pursuit of its space activities. It is therefore essential that ESA Management understands and accepts these risks, and that the risks are minimised. ESA has consequently established a safety policy covering its activities.

LEVEL I

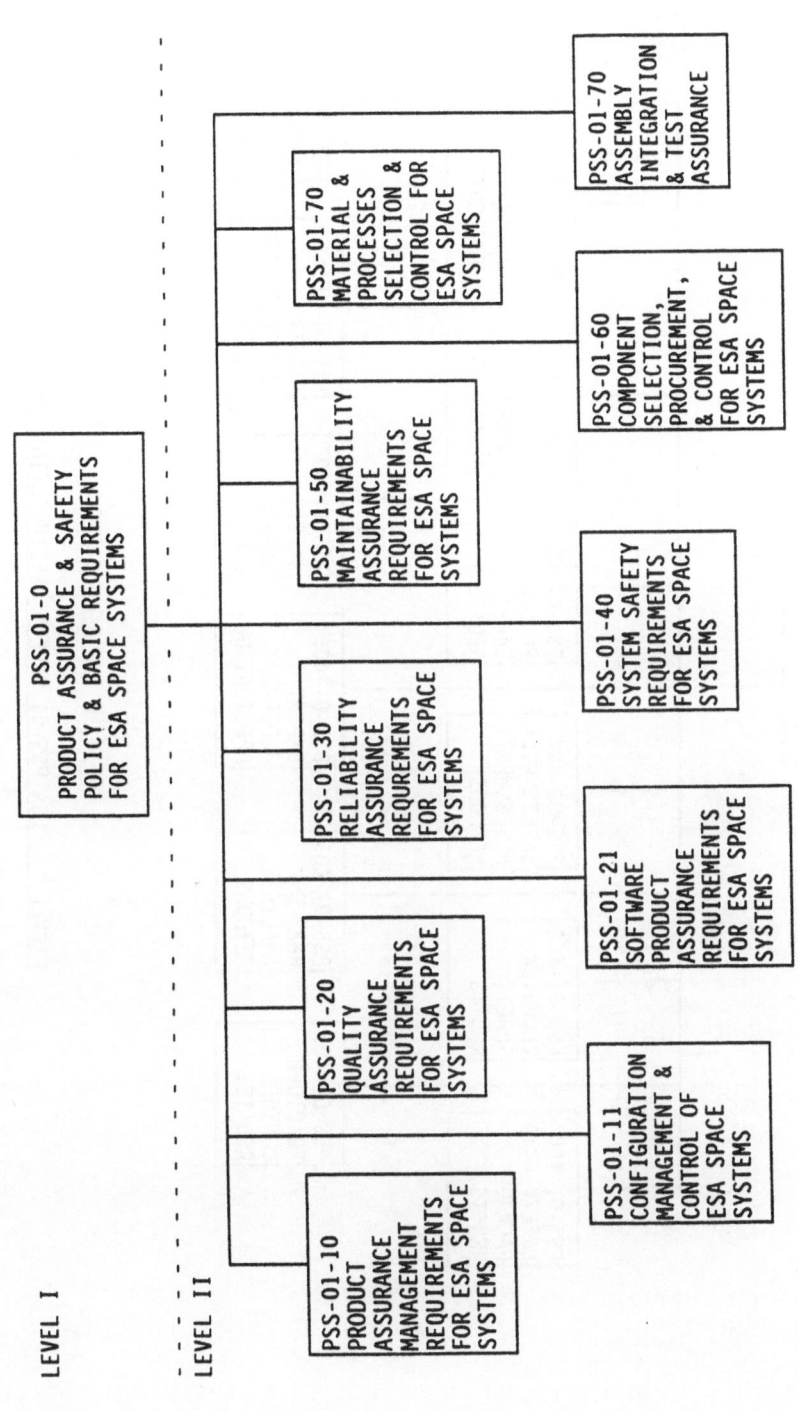

LEVEL II

FIGURE 1 - ESA PRODUCT ASSURANCE SPECIFICATION TREE, LEVELS I AND II.

19

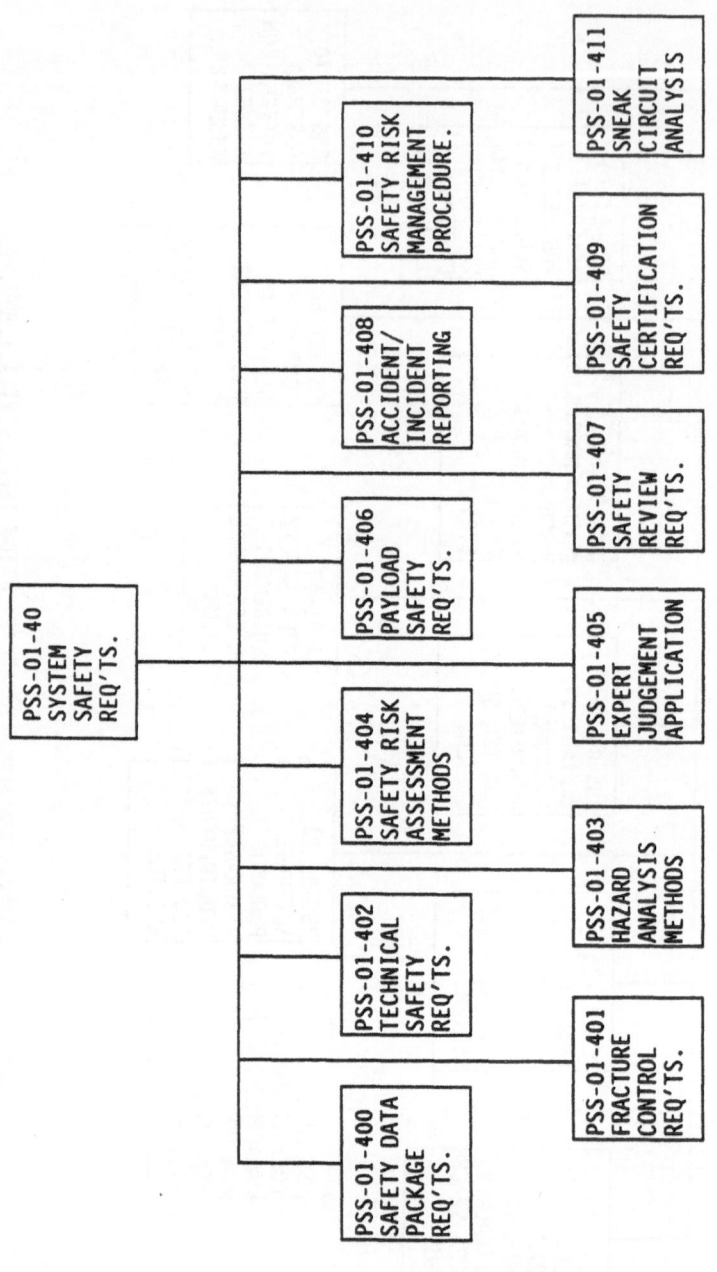

FIGURE 2 - ESA PSS-01 SAFETY SPECIFICATION TREE.

20

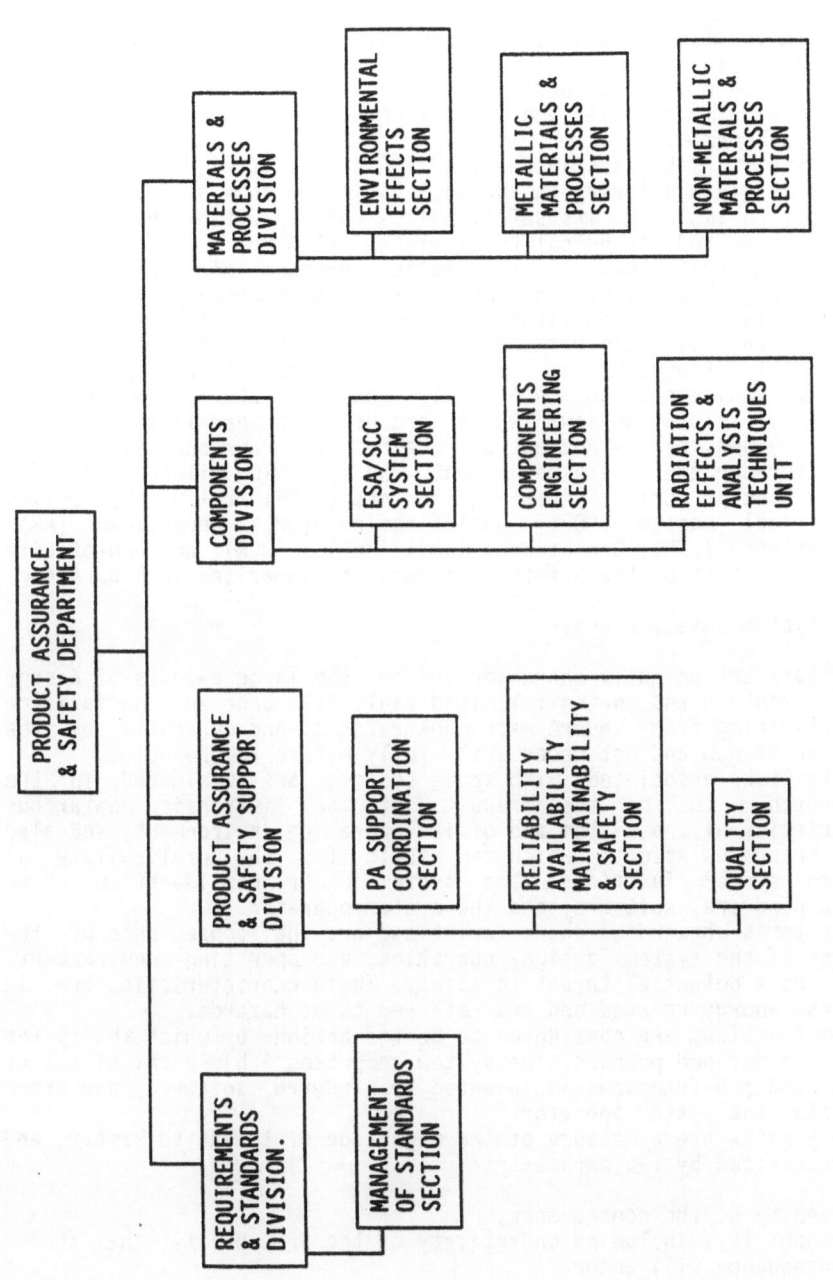

FIGURE 3 – ESA PRODUCT ASSURANCE AND SAFETY DEPARTMENT ORGANISATION.

21

The ESA Safety Policy is pro-active, and is intended to protect:
- human life;
- public and private property;
- investment (facilities and spaceflight hardware); and
- the environment,

from the safety risks associated with ESA's activities. Protection of human life is given priority in the application of this policy.

In order to implement its safety policy ESA has established a set of safety programme and technical requirements [ESA 88]. The ESA safety programme requires that a systematic process of safety risk identification, evaluation, and management is implemented. This process is based primarily on qualitative, deterministic analysis which is supported by probabilistic risk assessment and risk management.

The Safety Programme is applied from project conception to end-of-life. Implementation is formally monitored at defined Project Milestones by Safety Reviews which are held in parallel with the Milestone Reviews. Safety Reviews are chaired by a representative of the Product Assurance and Safety Department. The following milestone reviews are held:- Concept (CR); System Requirements (SRR); Preliminary Design (PDR); Critical Design (CDR); Qualification (QR); Acceptance (AR); Flight Readiness (FRR); Operational Certification (OCR); and End-of-Life (EOLR). The content of the Safety Programme is summarised in Figure 4.

4. Space System Safety Risks.

Space systems are uniquely characterised by: the large amounts of energy which they contain and control; limited fault tolerance and performance margins resulting from severe mass constraints; and currently, by the application of new and not necessarily fully mature technologies.

Safety risks associated with space systems are considered, in the ESA approach, to be the result of the intrinsic hazardous characteristics of the system and of its operating environment, and also from the hazardous effects which can result from the unreliability of the space system functions. The system under consideration here comprises hardware, software, and the system operator.

The system's hazardous characteristics are the consequence of the properties of the system design, operation, and operating environment, which can be a potential threat to safety. These characteristics are, in most cases, energy related and are referred to as hazards.

System functions are considered to be the actions by which the system achieves its defined purpose. The system comprises a hierarchical set of functions and sub-functions implemented by hardware, software, and where appropriate, the system operator.

Safety risks are a measure of the magnitude of threat to safety, and are characterised by two parameters:

- severity of the consequence;
- probability, including uncertainty of the probability, that the consequence will occur.

The severity of consequence is categorised as follows:

I   CATASTROPHIC:
    - loss of life, life-threatening or permanently disabling injury or occupational illness.

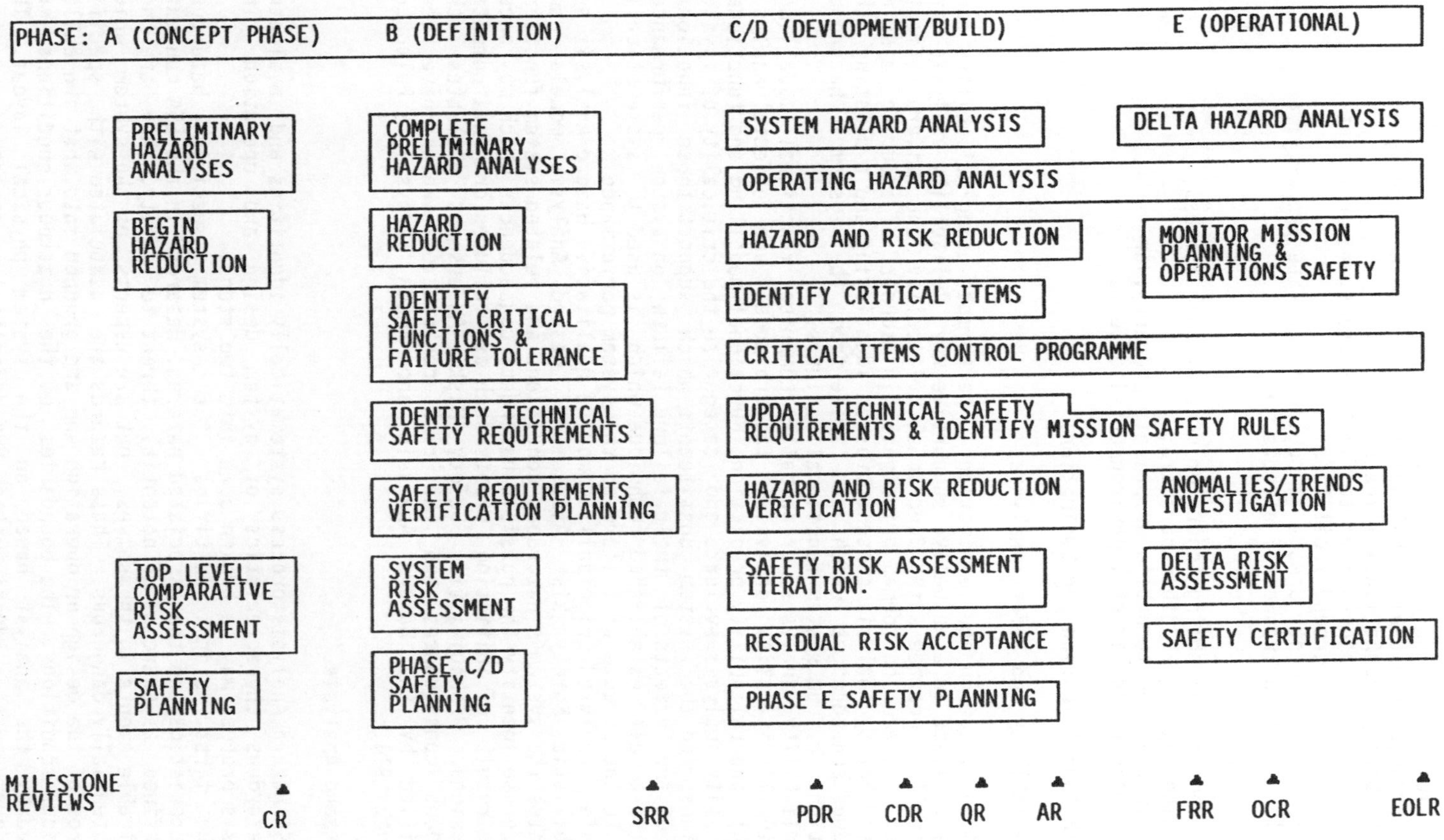

PHASE: A (CONCEPT PHASE)   B (DEFINITION)   C/D (DEVELOPMENT/BUILD)   E (OPERATIONAL)

Phase A (Concept Phase):
- PRELIMINARY HAZARD ANALYSES
- BEGIN HAZARD REDUCTION
- TOP LEVEL COMPARATIVE RISK ASSESSMENT
- SAFETY PLANNING

Phase B (Definition):
- COMPLETE PRELIMINARY HAZARD ANALYSES
- HAZARD REDUCTION
- IDENTIFY SAFETY CRITICAL FUNCTIONS & FAILURE TOLERANCE
- IDENTIFY TECHNICAL SAFETY REQUIREMENTS
- SAFETY REQUIREMENTS VERIFICATION PLANNING
- SYSTEM RISK ASSESSMENT
- PHASE C/D SAFETY PLANNING

Phase C/D (Development/Build):
- SYSTEM HAZARD ANALYSIS
- OPERATING HAZARD ANALYSIS
- HAZARD AND RISK REDUCTION
- IDENTIFY CRITICAL ITEMS
- CRITICAL ITEMS CONTROL PROGRAMME
- UPDATE TECHNICAL SAFETY REQUIREMENTS & IDENTIFY MISSION SAFETY RULES
- HAZARD AND RISK REDUCTION VERIFICATION
- SAFETY RISK ASSESSMENT ITERATION.
- RESIDUAL RISK ACCEPTANCE
- PHASE E SAFETY PLANNING

Phase E (Operational):
- DELTA HAZARD ANALYSIS
- MONITOR MISSION PLANNING & OPERATIONS SAFETY
- ANOMALIES/TRENDS INVESTIGATION
- DELTA RISK ASSESSMENT
- SAFETY CERTIFICATION

MILESTONE REVIEWS

CR   SRR   PDR   CDR   QR   AR   FRR   OCR   EOLR

FIGURE 4 - SAFETY PROGRAMME OVERVIEW.

23

II CRITICAL:
- temporarily disabling but not life-threatening injury, or temporary occupational illness;
- loss of, or major damage to flight systems, major flight-system elements, or ground facilities;
- loss of, or major damage to public or private property; or
- long term environmental effects.

III MARGINAL:
- minor non-disabling injury or occupational illness;
- minor damage to other hardware;
- minor damage to public or private property; or
- temporary detrimental environmental effects.

5. Risk Identification and Evaluation.

5.1 Reliability and Safety Analysis.

Risk identification and evaluation is performed using Reliability and Safety analyses in a mutually supportive way. Reliability analysis are principally concerned with functional criticality and failure effects aspects, and are performed principally in support of assuring mission success. Safety Analysis concentrates on identifying and evaluating the hazardous consequences which can arise from the system's hazardous conditions or from system functional failures.

Reliability analyses principally comprise functional modelling, functional criticality analysis, failure modes and effects analysis (FMECA), and reliability prediction. Functional modelling and functional criticality analysis evaluate and categorise the criticality of system functions, and the system constituents which support those functions, based on the effects of their failure failure on system performance. FMECA is a well established technique which is used to systematically evaluate the effects of failure on the system performance.

Safety analyses principally comprise qualitative Hazard Analysis and probabilistic Safety Risk Assessment. Hazard Analysis models and evaluates the system's hazardous conditions and related system functions in order to identify the possible hazardous consequences which may occur as the result of initiating events such as failures or environmental aggressions. Probabilistic Safety Risk Assessment evaluates the hazardous consequences and the risk contributors which have been identified by the Hazard Analysis and FMECA [Preyssl, Peltonen, Panicucci 91].

5.2 Hazard Analysis.

The ESA Hazard Analysis process systematically identifies and evaluates the hazardous characteristics of system design and operation. The analysis process may be broken down into two steps.

The first step identifies the system specific hazardous characteristics and the associated hazards. Hazards, in the ESA context, are defined as sources of potential threat to safety. They are not undesirable events in themselves, but are aspects of the system which are potentially dangerous. These hazards are associated with specific features of the design or operation and are grouped into what are called hazardous conditions. The boundaries of the hazardous conditions are defined by the analyst based on the system physical layout. The hazardous condition identification and definition is performed in order to derive a safety model of the system which represents its physical and functional arrangement. The safety model is documented in the Hazardous Condition Reports. These Hazardous Condition Reports identify relevant technical information, which includes: specific details of the hazards associated with the design feature(s) or operation to be analysed; the functions and operational phases which are associated with the design features or operation, their location in the system, and the operational environment.

In the second step, the safety model of the system, as documented in the Hazardous Condition Reports, is further evaluated by a qualitative analysis process which is similar to FMECA. This process derives the possible accident causes and accident scenarios associated with each hazardous condition, considering both physical and functional propagation. The scenarios culminate in a qualitatively ranked set of possible consequences of differing severities linked through the accident scenarios to initiating events and hazardous conditions.

Scenarios with common consequences and severities are grouped and documented in Hazard Consequence Reports which initially identify the related hazardous conditions, scenario initiating events, and the relevant scenarios. These reports are used as inputs for Safety Risk Assessment, and provide for the subsequent tracking, verification, and acceptance of hazard and risk reduction measures.

There are thus two principle outputs from ESA hazard analysis which are the basis for risk reduction, the Hazardous Condition Reports and Hazard Consequence Reports.

5.3 Reliability and Safety Analysis Interfaces.

The reliability analyses support the safety analysis through:

- the identification of functions and failure effects which are critical to safety;
- the identification of accident causes which are related to hazardous conditions; and
- provision of reliability predictions for the system functions which are critical to safety.

Figure 5 provides an overview of this reliability to safety analysis interface.

5.4 Probabilistic Safety Risk Assessment.

The ESA Probabilistic Risk Assessment (PRA) methodology constitutes a corner stone for risk reduction which is controlled by the Safety Risk Management process. This assessment identifies the significant risk contributors, and evaluates safety risk trends. It also provides for risk comparison in support of design trades and evaluates the potential for risk reduction through sensitivity analysis of basic initiating event probabilities. Probabilistic Safety Risk Assessment is therefore fundamental to Safety Risk Management as it supports assignment of project resources to risk reduction, while providing a relative measure of risk in support of trade-off and risk acceptance.

The Hazard Consequence Reports documenting the identified accident scenarios are used as the starting point in the risk assessment. Scenarios with the same degree of consequence severity are grouped together in tree structures called consequence trees. Each tree uses Boolean logic to trace a top consequence to its basic initiating events and the related hazardous conditions. For each consequence tree, the minimal cut set equation is determined.

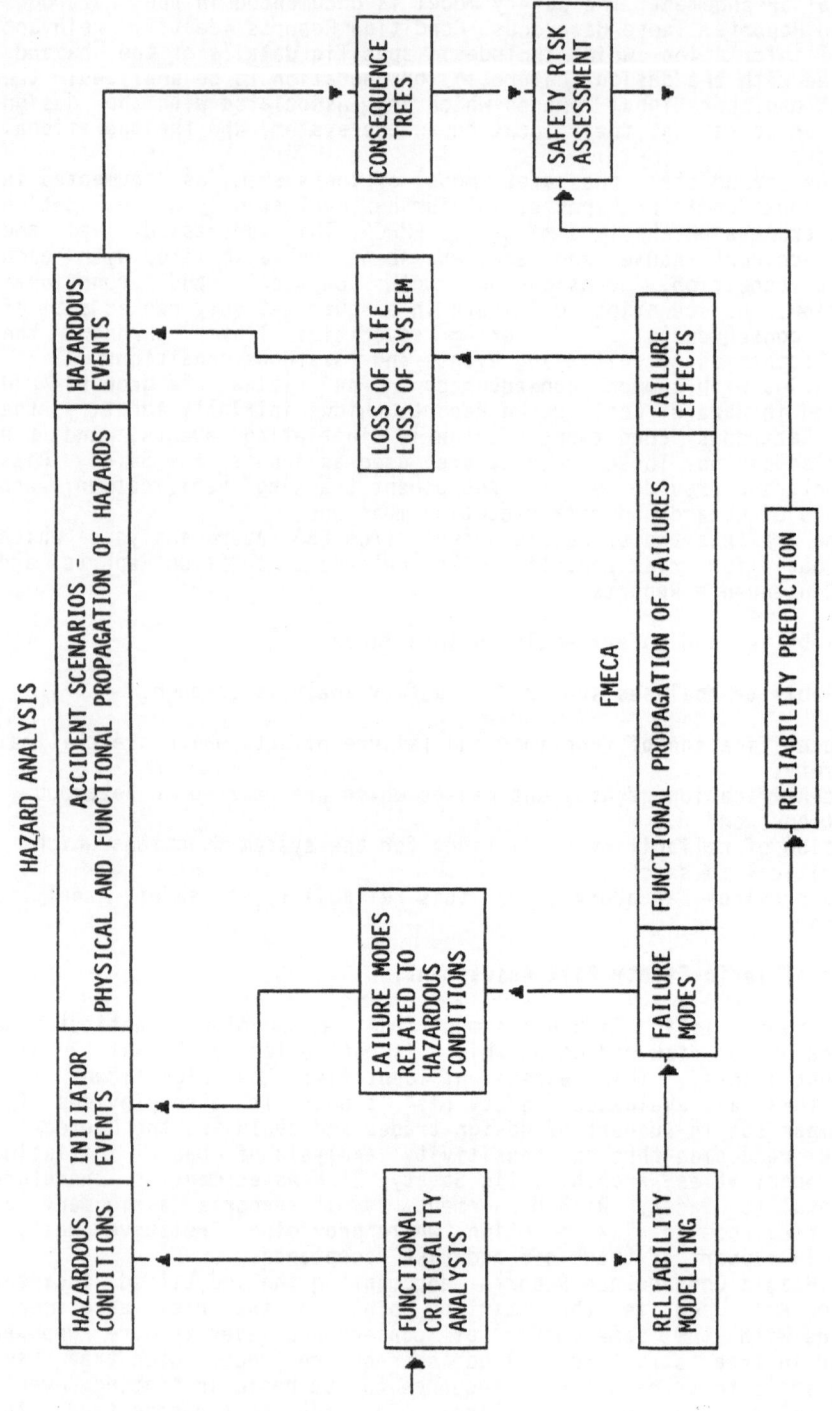

FIGURE 5 - RELIABILITY ANALYSIS SUPPORT TO SAFETY ANALYSIS.

26

The next step involves the determination of the probabilities, with their uncertainties, of all basic initiating events in the form of probability distribution functions. Typically in space applications, the complete spectrum of data sources from statistical tests to expert judgement has to be used to be able to cover all consequence tree basic events. The use of objective, i.e. directly relevant statistical data is, of course, favoured over data derived from similarity considerations or subjective data based on expert judgement.

Once all basic event probabilities are determined for a consequence tree the risk value can be computed. The top consequence probability is obtained by the classical approach of combining the basic event probabilities according to the minimal cut set equation. However, for the mathematical combination of probabilities, dependent uncertainty analysis is used. This analysis is such that the data sources determine the coupling of uncertainties and hence the uncertainty in the risk value. In addition, a quantitative subjectivity value is calculated for the top-consequence probability estimate based on the data subjectivities of the contributing basic events.

The risk value is displayed either as a probability interval or as potentiality. The probability interval is defined as the 90% confidence range of the probability distribution function associated with the risk value. The potentiality is a representative point-value in the upper part of the distribution (which is the most important part for risk assessment) while it is not as sensitive to the extreme tails, as the mean. Potentiality values are therefore well-suited particularly for risk contributor ranking and trade-off comparisons.

Safety risk assessment activity supports safety optimisation and risk management in several respects. In particular, PRA is used:

- to provide a method for the systematic, quantitative assessment of the risk to a system from the system point of view, specifying the uncertainty associated with that assessment;
- to drive the design from a safety perspective by identifying priorities for improvement as well as by highlighting the need for any additional investigations;
- to support the establishment of a balanced design based on a proper trade-off between safety and cost, by providing a basis for comparison of associated risk;
- to determine and evaluate the residual risk, and compare this with safety goals;
- to display the risk trend;
- to support the formulation and acceptance of the residual risk (quantitative safety targets and risk reference values).

6.0 SAFETY RISK MANAGEMENT.

The Safety Risk Management process comprises:

- Hazard and Risk Reduction;
- Hazard and Risk Control Verification; and
- Residual Hazard and Risk Acceptance.

6.1 Hazard and Risk Reduction.

6.1.1 Principles of Hazard and Risk Reduction.

The hazard and risk reduction process is based on a well established principle which has been applied in the qualitative risk management of space and military systems in the USA. It is applied to the payloads of the NASA Space Shuttle and was also applied during the development programme of the ESA Spacelab. It has been adopted by ESA for its safety programmes in a modified form, where it is supported by probabilistic risk assessment. It is defined in ESA PSS-01-40 and comprises:

a) Hazard Elimination;
b) Hazard Minimisation;
c) Control of Residual Risks; together with
d) Implementation of Warning and Safety Devices; and
e) Safe Haven and Escape and Rescue provisions.

The objective of hazard elimination and minimisation is to obtain the least hazardous system design and operational characteristics consistent with project objectives. Control of residual risks is implemented in order to minimise the probability of the residual hazards associated the hazardous conditions propagating to hazardous consequences. Controls are also applied to functions with hazardous failure consequences.

Warning and safety devices, together with the provision of safe haven and escape and rescue capabilities, provide the additional protection for personnel. This is necessary due to the fact that:

- manned spaceflight remains a relatively immature and risky undertaking;
- safety analysis is performed on a model of the system;
- the fidelity of model is limited by the analysts' knowledge of the system and by the ability to model the system based on that knowledge; and
- errors and omissions will occur in safety critical decision making during the life of the project.

Figure 6 gives an overview of the application of the hazard and risk reduction process to the hazard and accident scenario. Although this process is in general a joint safety and engineering activity, safety risk management is necessary in order to support the achievement and verification of optimised solutions.

6.1.2 Hazard Elimination and Minimisation.

As stated in paragraph 5.2, hazards are manifested as "hazardous conditions", which may comprise one or more hazards. These are usually associated with a specific design or operational feature of the system. Hazardous conditions, together with their associated hazards, are initially addressed by hazard elimination followed by hazard minimisation.

The application of hazard elimination and minimisation is of course practically constrained by the project and mission objectives. Hazards are usually associated with essential aspects of the system or its operating environment such as propellants, rocket motors, fuel cells, power distribution, batteries, pressure vessels, meteoroids and space debris, ionising radiation, solar thermal flux, etc.

Total elimination of all hazards is of course impossible and would have the effect of precluding the achievement of the project and mission objectives. Hazards can therefore only be eliminated in limited cases. However hazard elimination, where implemented, can result in a significant reduction of system risk. This also has a beneficial effect on the life cycle cost of the system, as hazards which are eliminated do

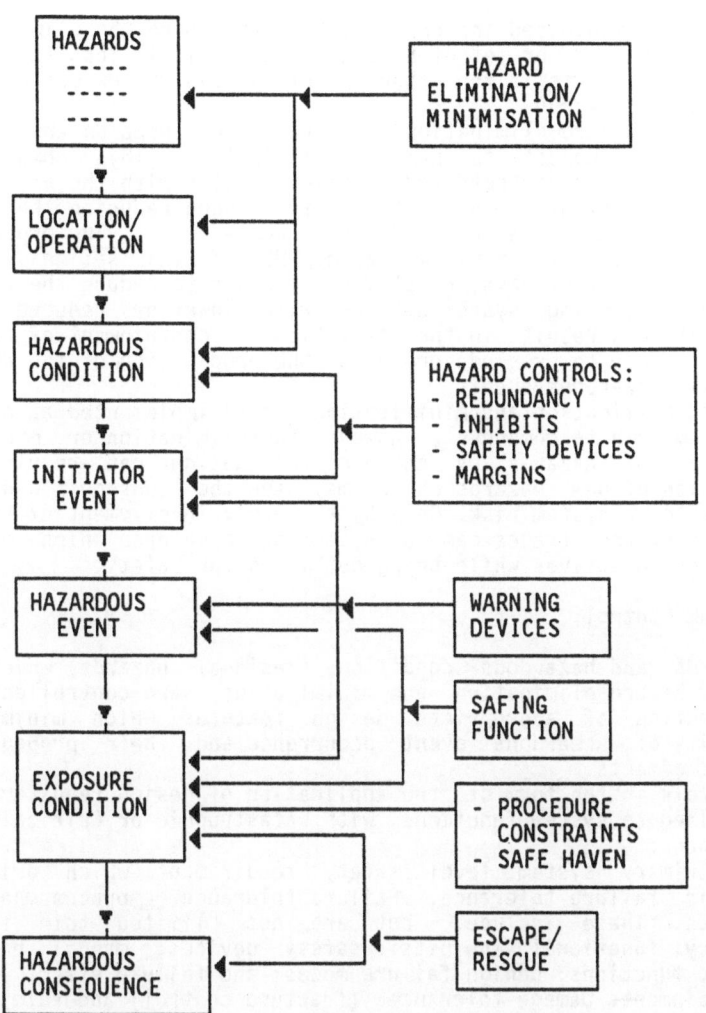

FIGURE 6 - HAZARD AND RISK REDUCTION PROCESS.

not have to be minimised and controlled. Furthermore the elimination of a particular hazard, or set of hazards, can result in less severe design and performance criteria for other aspects of the system design such as crew escape and rescue.

Following hazard elimination, the remaining hazards and hazardous conditions are subject to hazard minimisation. This comprises the selection of design concepts and characteristics with the objective of minimising the potential threat to safety through reducing the severity of potential hazardous events and their consequences. The process also known as designing for minimum hazard. Hazard minimisation drives the design in a number of ways. Firstly it attempts to reduce the risk from hazards by selecting system options with lower or reduced hazards. Secondly it can result in the isolation or containment of remaining hazards in order to preclude or reduce the propagation of the effects of potentially hazardous events.

Hazard elimination and minimisation can be implemented at all levels from system down to component, however the elimination or minimisation of one set of hazards by changing the design can result in the introduction of new hazards which may have the consequence of either higher or lower system risk. Only by the early involvement of safety in systematic system trades can a system be developed which meets its performance objectives while being optimised for safety.

6.1.3 Risk Control.

The hazards and hazardous conditions (residual hazards) which remain following hazard elimination and minimisation, are controlled through the selection of appropriate design features which minimise the probability of hazardous event occurrence and their propagation to hazardous effects.

Controls in the form of the application of design requirements are also applied to system functions with Catastrophic or Critical failure consequences.

The primary system level safety requirement which drives risk control is failure tolerance. Failure tolerance embraces many design approaches. These include, but are not limited to: functional redundancy; functional inhibits; safety devices; manual back-up to automatic functions; benign failure modes; and failure effect isolation and containment. Damage tolerance (fracture control) and safety factor requirements are applied to structures as an equivalent to functional failure tolerance, although structural failure tolerance is acceptable where it can be implemented. Common cause and common mode failure mechanisms are taken into account in the implementation of Failure Tolerance.

Performance margins are applied in support of failure tolerance. Margins include electrical and electronic component de-rating, use of conservative life times, selection of conservative system and equipment operational parameters. These are verified by qualification testing on flight standard products.

Application of safety requirements is based on the severity of potential hazardous consequences identified through hazard analysis and FMECA. Support is provided by the iterative application of safety risk assessment during all project phases. This supports the focusing of design and development attention on system elements with high risk contribution, and therefore high risk reduction potential. In addition, it identifies where concentration of testing and verification effort can result in the most effective improvement in confidence of the risk estimate.

6.1.4 Technical Safety Requirements.

The practical implementation of hazard and risk reduction is based primarily on lessons learned during the last thirty-five years of spaceflight. It is applied at different levels of design from system down to component level as the project progresses from concept to detailed design. They cover all aspects of the hazard and risk reduction process.

The lessons which have been learned have been developed into a myriad of safety and reliability design requirements and guidelines. Many of these are currently defined in numerous technical specifications. They are applied deterministically based on the results of the safety analyses and by the application of the hazard and risk reduction process.

For ESA Space Systems, generic system level technical safety requirements are defined in PSS-01-40. These requirements are intended to scope the largest and most complex systems while specifying the minimum standards which have been established as acceptable. They are based on extensive research of past space programmes. They can be tailored, or added to, in order to meet the specific needs of individual projects.

The requirements principally comprise the application of: failure tolerance; structural fault tolerance (fracture control); positive structural margins; performance margins, crew escape and rescue; and hazardous event detection, annunciation, and safing.

Detailed technical requirements and guidelines are given in ESA PSS-01-402, which is currently in preparation. This will ultimately form the basis of a safety engineering database. This database is intended to provide support to the safety analyst and the design engineer in the performance of hazard and risk reduction. It will also include the rationale behind the requirements, together with the associated hazards and possible related hazardous events.

6.2 Hazard and Risk Control Verification.

Hazard and risk reduction tracking is a fundamental part of safety risk management. It is the mechanism for assuring that the hazard and risk controls which have been identified as applicable are implemented and verified for effectivity. Where problems arise in implementation, it provides a means of identification of noncompliances with hazard and risk reduction criteria for relevant management attention. Figure 7 shows the logical flow of the hazard and risk reduction activities. This process is supported by the other Product Assurance disciplines including: Quality Assurance; Configuration Management and Control; and Parts, Materials, and Processes procurement and application control.

Prioritisation is applied based on the results of the Safety Analyses, and is initially based on the qualitative ranking of the Consequence Severity.

The Hazard Consequence Report (which can be considered as a residual risk report) is used to identify the applicable safety requirements based primarily on the consequence severity. The specific means by which these requirements are implemented are defined by safety and engineering personnel and documented on the relevant Hazardous Condition Report(s). This may involve selecting the appropriate specific safety requirement or safety design feature.

If for some practical reason a suitable and compliant solution cannot be identified for implementation, the relevant report is identified as potentially unresolved. This will then receive a higher

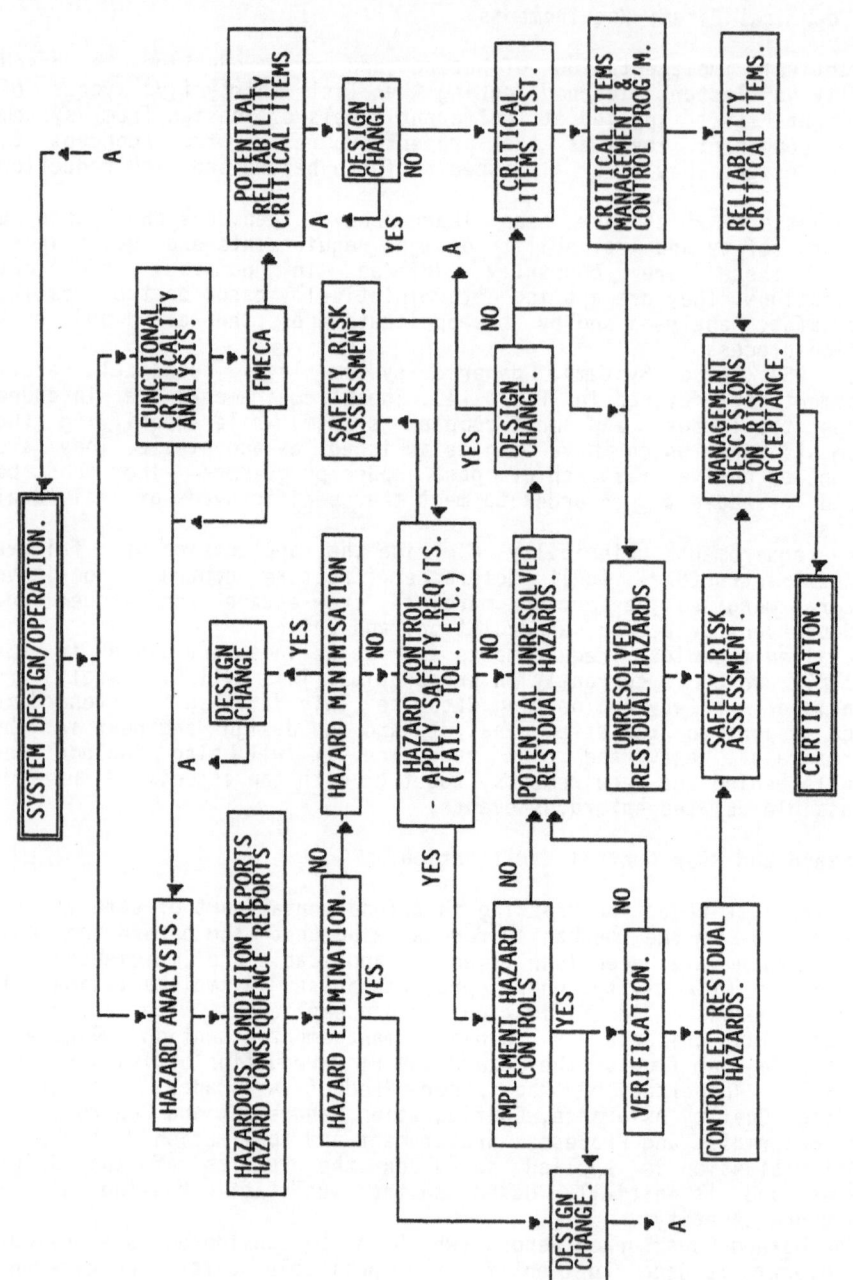

FIGURE 7 - HAZARD AND RISK REDUCTION IMPLEMENTATION.

level of management attention than reports where hazard and risk reduction implementation is in compliance with the applicable requirements. If one or more noncompliances with safety requirements are found to exist where Catastrophic or Critical consequences are possible, then the associated item is identified as safety critical. One or more critical items may therefore be associated with one unresolved Hazard Consequence Report, or Hazardous Condition.

Identified Safety Critical Items are subject to a Critical Items Control Programme. The programme provides for an elevated level of Management, Engineering, and Product Assurance attention for those items identified as critical. The items are ranked in order of criticality in order to support prioritisation for attention and resource assignment.

Through the Critical Items Control Programme, safety critical items may be eliminated wherever this is technically feasible, or where this is not possible, they are subject to strengthened design, manufacturing, and qualification controls which address the safety critical aspects of the each item, so reducing the resultant risk. Probabilistic risk assessment supports this process by evaluating the effect of the measures taken on the risk probability so supporting the ranking process.

Where the relevant safety requirements can be complied with, the Hazard Consequence Report continues to be processed in a nominal manner. If, for some reason, a defined safety requirement cannot be satisfactorily implemented, or its implementation cannot be verified as complying with the intent of the requirement, then once again the relevant report is identified as potentially unresolved. Furthermore, if risk assessment determines that, in spite of compliance with defined requirements, significant risk contributors remain in the design, then the relevant report is also identified as potentially unresolved. At this stage, additional safety critical items may also be identified.

Design changes, manufacturing nonconformances, failures, and waivers are also be evaluated for impact on analysis validity or hazard control implementation or verification.

6.3 Prioritisation, Resource Allocation and Risk Influencing Factors.

It is essential that the safety optimisation effort is managed so that the limited project resources are used to the greatest effect. This necessitates that a system of ranking be applied to the system risk contributors. A process of qualitative and quantitative risk contributor ranking as outlined in the above paragraphs is currently implemented. However this safety analysis based ranking alone does not provide a sufficiently comprehensive basis for resource allocation and decision making. Due consideration needs to be given to the additional "real world" characteristics of the risk contributors which are often over-looked in conventional safety analyses.

Both the safety and the supporting reliability analyses are focused on the identification, and qualitative and quantitative evaluation, of risk contributors in the conceptual, design and development phases. This lays the foundations for risk reduction through risk management. However, the impact of quality control and project organisational aspects on the overall safety and reliability risk, needs to be taken into account. A good system design which, according to the safety and reliability analyses, represents an acceptable risk, can be severely degraded by defective management or quality assurance during the manufacturing, assembly, and test stages. Similarly, in the operational phase, lack of awareness of the full implications of in-flight anomalies and system operating boundaries can ultimately

33

result in exceeding qualified safe operational limits.

This is where the conventional safety and reliability analyses tend to be insufficient. They are not capable of efficiently reflecting the quality related information for an up-to-date presentation of the safety risk imposed by the risk contributors of the system configuration.

In order to enhance the use of risk analysis results as a safety optimisation aid, and to extend prioritisation and resource allocation to cover also impact of manufacturing and operational factors, it has been proposed [Wright & Peltonen 92] that the dynamic impact of these quality related "Risk Influencing Factors" should be considered in support of resource allocation during Safety Risk Management.

Typical Risk Influencing Factors include:

- open space qualification status of Parts, Materials and Processes;
- open major nonconformances;
- critical manufacturing processes;
- part, material and process alerts;
- critical assembly, integration, test and operation processes;
- unsatisfactory Product Assurance audit results;
- test failures;
- in-flight anomalies.
This proposal is currently under further evaluation.

6.4 Residual Risk Acceptance.

For large projects, effective risk acceptance is complicated by:

- the multiple tier project industrial organisation;
- the many residual hazards and hazardous conditions; and
- the influences of waiver, failures, nonconformances and critical items.
A hierarchical system of safety risk acceptance is therefore necessary. This acceptance system is based on the formal evaluation and acceptance of:

- controlled residual hazards;
- unresolved residual hazards;
- identification and control of safety critical items.

The level of management which is responsible for safety risk acceptance is based on the qualitative ranking of the risk being accepted. Ranking is achieved by taking into account hazard consequence severity categories, whether the hazard is characterised as unresolved or controlled, and the number of safety critical items associated with the risk being accepted.

Safety risk acceptance is implemented incrementally starting with hazard and risk reduction. The acceptance stages which coincide with the major programme reviews, comprise the acceptance of:

- the defined applicable system level technical safety requirements;
- the identified specific and detailed hazard minimisation and control requirements;
- the identified verification methods;
- the evidence of verification of implementation of the defined hazard controls together with acceptable levels of safety risk.
As can be seen, each acceptance stage builds on the completion of the previous acceptance stages. Acceptance only remains valid if the

design standard complies with requirements, and the build standard complies with the design standard, or where deviations occur these are evaluated for impact to safety and any effect taken into account in the safety analysis.

7. Summary and Final Remarks.

This paper has presented ESA's current system safety approach and some proposals for its improvement which are under consideration.

ESA's safety policy is one of pro-active protection of life, investment, property, and the environment. The policy is implemented through the minimisation and formal acceptance of the risks associated with the system and its operation. Risk minimisation is achieved by optimising of the system design and operational characteristics for safety, within the constraints imposed by the project objectives.

Effective safety optimisation in the space project environment is achieved by systematically identifying the risks and risk contributors associated with the system under development, and by the timely concentration of the available engineering and safety assurance resources on reducing those risks. System risks are considered to be the consequences of the hazardous characteristics of the system design, operation, and operating environment, together with the unreliability of the system's functions. The system is considered to comprise hardware, software, and the operator.

A combination of qualitative and quantitative analysis and ranking is applied in the risk reduction, management and acceptance process as a support to the allocation of project resources to the system safety process.

In spite of the strict implementation of the system safety process, spaceflight will remain a high risk activity for the foreseeable future. ESA's safety programme, methods and technical requirements are therefore subject to a process of continual improvement in both technical effectivity and cost effectiveness. A important part of this improvement process is an initiative taken by ESA in 1991 towards the standardisation across Europe of space system Product Assurance and Safety and technical requirements based on the ESA PSS series of specifications. The objective of this initiative is involve the National Space Agencies in the continued development of these specifications, which are intended to be adopted for all European Space activities. In support of expanding the scope of this standardisation initiative, a series of meetings are being held with NASA, with the objective of achieving equivalence and mutual acceptance of each others standards. A similar activity has recently been started with the CIS (formally the USSR).

REFERENCES:

ESA: System Safety Requirements for ESA Space Systems and associated equipment, ESA PSS-01-40, Issue 2, September 1988.

Preyssl, Peltonen, Panicucci: Safety Risk Assessment for ESA Space Programmes. Proceedings of the Space Product Assurance for Europe in the 1990s Symposium, ESA SP-316, August 1991.

Wright, Peltonen: Safety Optimisation through Risk Management. 43rd Congress of the International Astronautical Federation IAA-92-0379, August 1992.

The Swedish State Railways' experience with n-version programmed systems.

Author

Johan F. Lindeberg, SINTEF DELAB

N-7034 Trondheim, Norway

Abstract:

This paper contains a resume of the Swedish experience with n-version programmed systems the last 12 years; ideas and attitudes, the growth of experience, current understanding, and future goals. The two control applications; Computer controlled Interlocking System (CCI) and Automatic Train Control (ATC) will be explained.

A number of ATC systems were made as single-version systems. A comparison between single- and dual-version designs will be given. The paper also contains a discussion about the discrepancies between what theory predicts and what experience teaches us about two-version programmed systems.

Introduction

Some twenty years ago The Swedish State Railways (SJ) decided to upgrade their railway control systems. SJ believed there would be a net gain both in terms of economics and safety by utilizing computer technology.
Fig. 1 illustrates the main items.

One great challenge was the development of an ultra reliable high capacity communication link between train and track for the ATC. Five different concepts were tested. Two of them were almost equivalent, but only one were selected. ATC version 1 was then developed and put into regular operation in 1981.

This gives SJ over 16 000 operating years of ATC experience. The ATC is applied on about 60% of the Swedish network, carrying more than 95% of the total traffic. The Swedish network is about 11000 km long.

ATC version 2 is now under development and is planed to go into operation in 1993.

The computerised interlocking system was developed and initial installation was done in 1969, today covering about 10% of the interlocking systems. SJ has now more than 2500 operating years experience with CCI.

A few computerised Central Train Control (CTC) systems were also developed and put into operations, giving altogether 200 operating years.
Fig. 2. illustrates the main technical components constituting the railway control system.

The total experience in terms of operating years is tabulated in table 1.

ATC

The ATC system is used as an extension of the signal system into the drivers cab. In Sweden only speed signalling is used. The Swedish ATC system is something halfway between ATP (Automatic Train Protection) and ATO (Automatic Train Operations).

For various reasons SJ decided to develop two variants of the ATC.

ATC variant 1 has a central unit consisting of three identical computers and the same software package copied into each of the three computers. (See fig. 3.)

ATC variant 2 has a central unit consisting of one computer with two different software packages executing in sequence. and comparing the results. (See fig. 4.)

Safety

SJ's courageous decision to introduce computers into their railway control systems was connected to their way of viewing the safety issue.

SJ claimed a substantial net safety gain by using computers even if it were beyond the state of the art to prove such systems fault free or fail safe [Sterner 81]. Calculations showed that high reliability/availability and low total cost were very important contributors to the overall safety level.

Stopping trains is no longer in itself an acceptable safe response to an equipment failure, because you then have to use exception procedures which in practice increases the local risk level orders of magnitudes.

High acquisition cost or other obvious costs may prevent an administration from investing at all, or spends too many years before the coverage is complete. During this time there is an unnecessary high risk level.

SJ has observed a net safety gain. A recent study reveals that administrations using ATC in average have a higher safety level [Hope 92]. The correlation is very high, but does not prove a causual relationship.

The effect of introducing computerised interlocking is not significant yet.

Comparison of ATC variant 1 and 2

In the seventies the main question was how large net safety gain ATC would offer.

One preliminary answer is given in a European railway survey published May 1992.

Both ATC variants had their strengths and weakness. It was therefore decided to accept then as equivalent for SJs purposes.

It turned out that the European railway administrations had, and have, different opinions over what attributes should have what weight. Therefore each administration have to make their own **ATC life cycle benefit/cost analysis**, or better, harmonise their views in order to make cross-acceptance possible. Table 2 contains a scheme for comparing the variants. Some of the most relevant attributes are indicated.

Fifteen years ago people were very enthusiastic about the benefits of diversity, To day they judge both cost and benefits.

One general technical criteria for judging a solution is its ability to function correctly in the presence of physical faults in hardware and electromagnetic interference, logical faults in hardware or software, and incorrect input.

Diversity

The diversity solution has in operations some ability to detect errors on different levels as indicated in Table 3, in average about 10-20% of remaining logical faults.

The diversity solution tends to have a lower availability because it shuts down or restarts from scratch on any discrepancy between the two channels:

- either because of encountering a logical fault or because of
- inherent weakness in the comparison algorithm.

Conclusion

Based on an operating experience equal to 16000 operating years SJ maintains:

- Use of ATC results in a safety gain. This is supported in [Hope 92].
- The one- and two-version ATC solutions are almost equal. No wrong side failures have been observed in either solution.
- It seems difficult to come in the vicinity of what the most optimistic theoretical models predict for two-version solutions.

A number of factors outside the single/dual aspect contributes significantly to the total safety. Availability is one of these.

Other railway administrations may have other operating conditions, investment policies, and attitudes.

They therefore have to make their own ATC life cycle benefit/cost analysis.

References

[Sterner 81] B. Sterner
              The Elusive Safety - The Problem of Software Evaluation
              ORE colloqvium Technical computer programs,
              Madrid 1981.

[Hope 92]   R. Hope
              Rational spending on safety brings results,
              Railway Gazette International, May 1992.

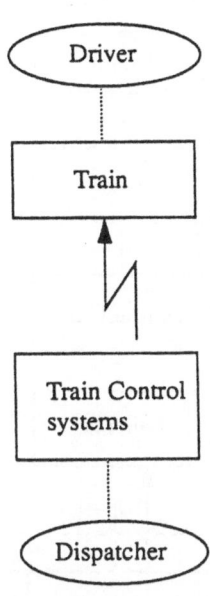

Figure 1 Train Control.

Table 1 : Operating experience.

Unit	No. of component units per 1993	No. of opr. yrs	Average No. of hard faults/yr	
ATC CU variant 1	1000	12000	470	
ATC CU variant 2	300	3600	100	
ATC transmission	1300	15600	350	
ATC transponder	40000	480000	200	
ATC encoder	17000	200000	200	
Interlocking	1600	2500		
CTC	40	200		

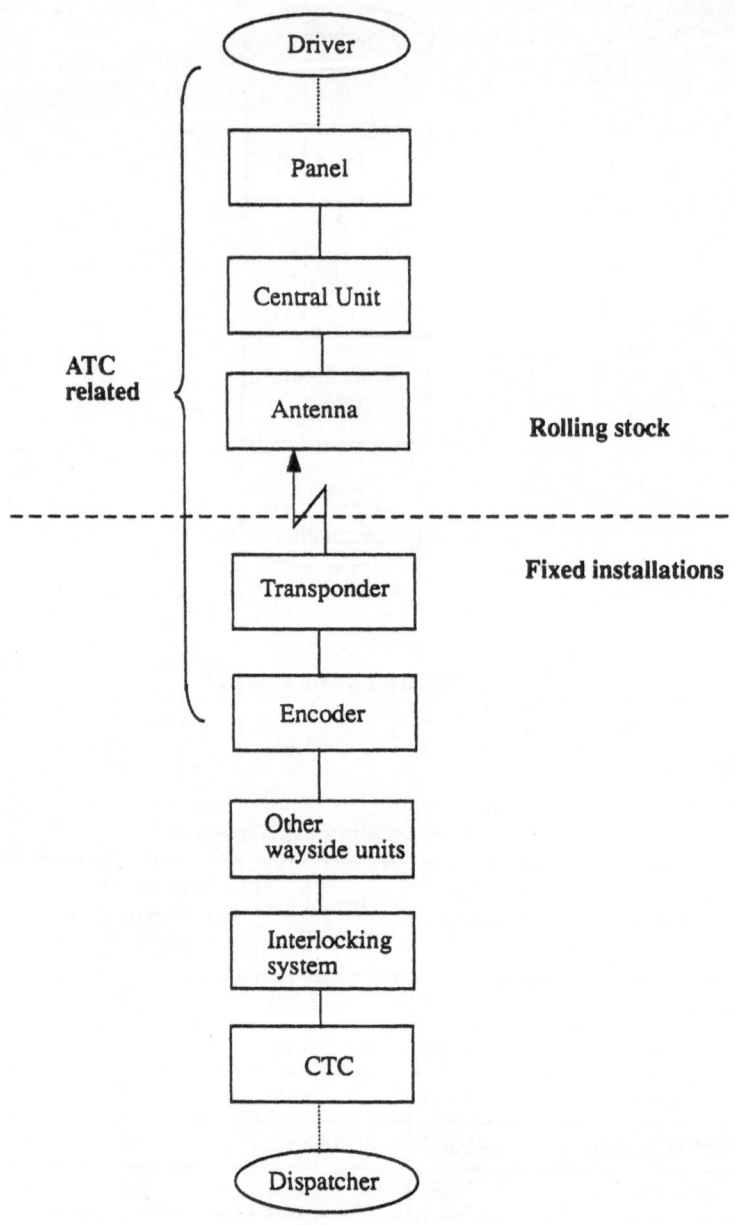

Figure 2 Block diagram of main components.

40

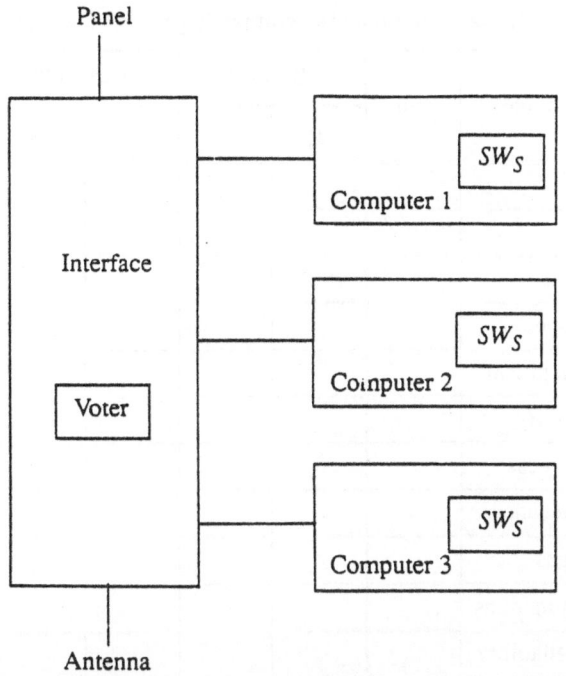

Figure 3 Central Unit for ATC variant 1.

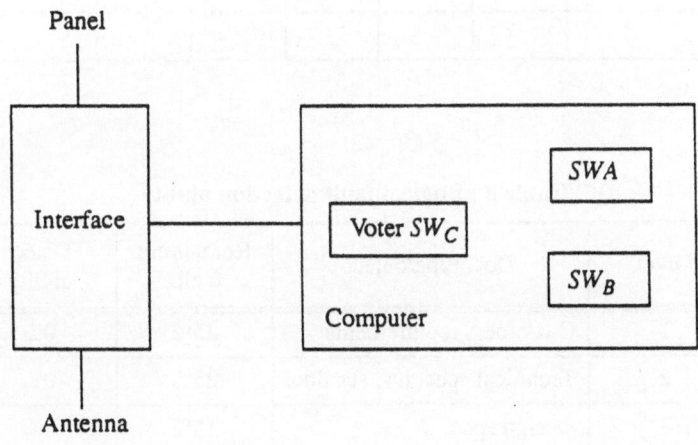

Figure 4 Central Unit for ATC variant 2.

Table 2 : Scheme for comparing the ATC variants.

Attribute	dim.	Variant 1		Variant 2		Weight
		qty	cost	qty	cost	
.						
.						
.						
Req. spec.						
Development						
Testability						
Robustness						
Maintenance						
Changes						
New functions						
Unavailability						
Unrelability						
Price						
.						
.						
.						

Table 3 : Logical fault detection ability

Level	Doc. type/object	Remaining faults	Detect ability
1	Customers requirements	25%	0%
2	Technical spec. hw. sw. doc.	25%	0%
3	Design spec.	15%	0%
4	Detail design	15%	20%
5	Hw.logic, code	20%	98%

Potential Difficulties in Managing Safety-Critical Computing Projects: a Sociological View

Margaret Tierney
Research Centre for Social Sciences
Edinburgh University
56 George Square
Edinburgh EH8 9JU

Abstract

The paper reviews the emergence of project management frameworks in commercial computing environments. It suggests that there is a tension between "scientific" and "industrial" solutions to controlling software development, and that most strategies for dealing with the management of the software labour process, and of relations with users and the rest of the organisation, are shaped by the latter set of concerns. In contrast, certain kinds of safety-critical software development depend heavily on a "scientific" commercial environment and labour process - formal methods, and their practitioners, being a case in point. After reviewing some of the work practices of formal methodists in high-integrity computing, the paper concludes with some research questions about the special difficulties facing project managers of safety-critical projects.

1 Introduction

In its basics, managing safety-critical projects is just an extension of normal practice. As with any computing project, the manager must construct budgets and schedules which justify and describe its cost, length and complexity; its order of priorities; and its allocation of particular people to particular tasks. To pursue the work, s/he must select from the available sources of expertise; configure staff into particular divisions of labour; and invoke methodologies for controlling and evaluating progress as the project unfolds. These are the staples of project management, regardless of the problem domain being addressed. Over the last two decades, a number of strategies have evolved with respect to all these elements, which offer (limited) means of achieving adequate project management. The question addressed in this paper is why implementing some of these established strategies may remain problematic for managers of safety-critical computing projects; why the organisation of safety-critical computing may be different in kind from mainstream commercial practice.

With this question in mind, there are two related issues I would like to explore. First, that most existing project management frameworks - with their attendant rationales and tactics - have been generated as solutions to the persistent commercial demands made on project managers: to maximise productivity, minimise development costs and, more recently, ensure that technical staff possess the expertise to meet users' requirements. In some contrast, managers of safety-critical computing projects must also satisfy a fourth and equally pressing demand: to maintain control over the safety and reliability of the thing produced. In this respect, those involved in this field face novel problems, where the organisational facets of good project management necessarily interwine with problems of demonstrating how trustworthiness is being achieved. We may well query whether the imperatives of demonstrating reliability ensure that safety-critical managers must reject certain tactics which are solely geared at controlling for cost or productivity.

Secondly, it is worth exploring whether the socio-economic constituency of high-integrity computing, by virtue of its own unique dynamics, confounds some of the assumptions underlying mainstream project management frameworks. The nexus of interactions found in this niche between the organisations who sponsor safety-critical projects, the specialist occupational groups who design them and the kinds of artefacts they produce, suggests that safety-critical computing, to some extent, marches to its own drum. Since safety-critical projects are embedded in a set of relationships which differ (in some respects, quite radically) to those found in mainstream practice, the organisation of development work in this niche may require unusual forms of project management.

In the first section of this paper, I will review the emergence of managerial frameworks for organising software development, to illustrate how they have evolved along two different (though overlapping) paths: the "scientific" and the "industrial". While the former is yielding many of the most promising means of obtaining reliable software, its focus remains in a state of tension with the commercial pressures and opportunities which have shaped "industrial" amendments to those innovations. In the next section, I will illustrate the effect the "science focus" has on shaping the organisation of software development in safety-critical projects (and, hence, on shaping how the staples of good project management may be tackled). Here, I will briefly look at the work environment of one group of specialists involved in high-integrity computing - formal methods practitioners. While a profile of the work practices of this particular group by no means describes the dynamics of safety-critical computing as a whole, a sketch of their work practices may illuminate some of the special labour and product market conditions found in the safety-critical niche. Finally, drawing on this sketch, I will raise some questions about how the dynamics of high-integrity computing may make it difficult for managers to account for their project's progress - its costs, scheduling, and so on - in standard "industrial" terms.

2 The emergence of project management frameworks

We can trace the explosion of managerial innovations to co-ordinate and control computing projects to the identification of "the software crisis" at the NATO/Garmisch conference of 1968 [Naur and Randell69, Pelaez88]. This crisis referred to the way in which software had become the dominant constraint on the take-up of computer-based systems despite rapidly-expanding markets [Friedman89]; software production lacked an intellectual and organisational framework within which detailed prescriptive technical design practices could be deployed, and out of which commercially-viable products could emerge. In acknowledgement of this crisis, the management of software (and systems) development has come in for sustained analysis and reform over the last two decades.

I will focus in this paper on software development practices. In part, this is because it was software which constituted the original "crisis" and, hence, was first made subject to systematic regulation and restructuring. In part, also, it is because software remains the single most complex component of a system; the urgency of establishing and maintaining a framework to control its development has hardly abated in two decades.[1] Finally, though, a focus on software reveals the historical cleavage between "scientific" and "industrial" approaches to its management - a cleavage which is pertinent to the management of safety-critical computing.

At the Garmisch conference of 1968, and subsequently, a variety of causes for the software bottle-neck were identified. Indeed, not least of the difficulties in proposing solutions, was that these causes were (and are) so diversified. The Garmisch participants were the inheritors of all the traumas of third-generation computing: the unbundling of software; the

---

[1] While there is no technical inevitability about concentrating systems' complexity in its software, it has been principally through software innovations that the capability and malleability of computer systems have increased - which compounds the problems of controlling for that additional complexity through the software.

slipping deadlines and escalating costs; the chronic shortages of skilled staff; the difficulties in co-ordinating information exchange in large projects; the gap between what users expected, and what producers could actually deliver; the disasters, or near-disasters, which could be traced to flawed software designs; the over-emphasis on coding and under-emphasis on design and specification; the failures in understanding and documenting the contents of programs [Friedman89, Naur and Randell69, Pelaez88].

Not all these problems were (or are) directly connected to each other, and the most successful management strategies which have emerged since the late 1960s have tended to address particular *types* of software production problems, as opposed to "silver bullet" cure-alls [Brooks86]. At the heart of the problem is the fact that though software is a commodity - designed, traded and used in commercial environments and thoroughly subject to the pressures of the market - it is a commodity with the special property of being "only slightly removed from pure thought-stuff" [Brooks75 4], the abstract product of mental labour. Software thus lends itself to rigorous scientific analysis in a way which other commodities do not.

Though there are, in practice, many over-laps between the "scientific" and the "industrial" management of software, solutions addressed at software production in its guise as abstract mental labour have been forged by the software scientists, ie. applied academics devising medium- to long-term technical strategies for gaining intellectual control over software's content. In contrast, software industrialists, ie. managers coping with the short- to medium-term problems of producing adequate software for immediate and direct use by others, have tended to devise organisational strategies for managing the processes of software development and implementation.

It seems likely that, in safety-critical projects, (aspects of) the scientific approach have shaped the development environment to which managers react, particularly with respect to meeting reliability requirements. Since these managers are also expected to retain normal "industrial" controls over the cost, pace and progress of the project, it seems likely that in the safety-critical arena, there is some tension between the two paradigms. To explore how these tensions might arise, I will first review some of the principles informing both approaches and then look at how they have been applied to two main clusters of innovations in software management practice over the last two decades.

2.1 Software engineering: managing software v. managing its commercial environment

A dominant argument which surfaced at the NATO/Garmisch conference, and which has pre-occupied theorists and practitioners since, is that unless software can be understood, it cannot be controlled. That is, its very structure and contents must be made intellectually visible; its abstract "thought-stuff" must be available for scrutiny by fellow technical staff, managers or users. Opening up the black-box of what software is and does, was seen as the first requirement for ensuring adequate management of its development and implementation. If software could be controlled "from within", by means of standardised methodological guidelines on how to conceive, design and write programs, then the problems attendant on software's complexity, functionality and quality could be redressed.

Software engineering - the term chosen in the Garmisch conference to express moves in this direction - involved "the need for software manufacture to be based on the types of theoretical foundations and practical disciplines, that are traditional in the established branches of engineering" [Naur and Randell69 13]. This new focus, it was hoped, could re-shape the esoteric and unsystematic "craft" of programming into the more intellectually visible discipline of "engineered" software production [Buxton90]. The forging and reworking of detailed software engineering techniques during the '70s and '80s - structured programming and structured design, software phases and life-cycles, standardisation of design and programming methods etc. - have, in one form or another, transformed the extent to which the structure and content of software can be made amenable to managerial direction and supervision [Sommerville89, Ince and Andrews90]. However, behind the very term software engineering, and the distinct and various strategies it has spawned, lies a deep ambiguity about the most appropriate means to solve software production problems [Pelaez88].

On the one hand, the genesis of detailed techniques to control the visibility, complexity, size and structure of software has been firmed rooted in the intellectual tradition of programming as a mathematical science [Dahl et al.72, Dijkstra69]. An assumption of this tradition is that if software design is properly informed by mathematics, then software will become more translucent, reliable and robust, and its management commensurately easier. In this conception, the science comes first and the benefits, whether commercial or managerial, follow. On the other hand, for most practitioners, software engineering has never been synonymous with science: its guiding principle is that managing software is inseparable from managing the occupational, organisational and marketing processes through which it is done [Macro and Buxton87, Friedman89]. This is less an anti-science position, than a pre-occupation with the "external" buffeting imposed on software production by, for example, staffing shortages, too much or too little communication between software producers, dismayed users, deadlines missed etc.

We might assume that the science tradition, as the minority voice in the history of software engineering, has had little impact on detailed project management practices in commercial computing. However, an examination of managerial innovations since the late 1960s does not bear this out. The contributions of software scientists have centred on ideas of visibility, rigour, simplicity and divisibility - ideas which have been applied to the technical management of software itself (eg. structured programming) and to the software labour process (eg. "scientific management" of the division of labour).

In practice, though, these contributions have had a varied history: some have been abandoned, most have been diluted or reworked. This is hardly surprising, for industrial project managers must continuously grapple with the economic and political environment within which software production is embedded. For example, simplicity may be a good design principle (especially where control over the software's reliability is of major importance), but the market generally favours complexity of design; profitable software is often premised on the availability of add-ons and enhancements [Pelaez88, Brady et al.92]. Or again, a very detailed and hierarchical division of software labour may provide for a more predictable and controllable labour process. However, it may also place severe strains on the manager's ability to retain or regroup software staff so as to generate new, and more pertinent, kinds of technical expertise to answer constantly changing requirements [Fleck and Tierney91, Friedman89, Tierney91].

Thus, in the forms of "industrial" software engineering which have come into common use in commercial computing practice, the focus is less on software production as a tightly-bounded and rigorously-delineated process, than on the dynamic and open-ended relationship between software and the environment which surrounds it. We can see this trend for "industrial" software engineering criteria to dominate "scientific" ones, when we look at the emergence of two major forms of managerial intervention in the software production process.

2.2 Structured programming, and its descendants

The development of structured programming techniques, particularly in the early post-Garmisch days, was firmly in the hands of academic and scientific programmers, rather than managers or management theorists [Friedman89]. Many of these scientists were members of IFIP Working Group 2.3 - a group who formed after the break-up of Working Group 2.1, to specialise in progressing programming methodologies [Pelaez88]. This group, which included people like Dijkstra, Hoare and Gries and which proved extremely influential in altering the orientation of software development, shared a common interest in raising the scientific level of computer programming. For them, many of software's failings stemmed from the chaos of its production, in frantic markets which hyped software beyond its capabilities; their intellectual preference was to turn their back on the market and reform programming fundamentals [Pelaez88]. Structured programming was the (series of) means suggested to structure complexity into intellectually manageable pieces (eg. limiting program size, avoiding GOTO statements, standardising practices to structure the flow of control through the program etc).

As a corollary of advances made in structured programming, the design process was also seen as being ripe for scientific analysis and reform. Techniques for structured design (eg. modular programming, functional decomposition, stepwise refinement, top-down development, information hiding etc) had, by the late 1970s, begun to filter into use in many large DP departments, aided by the availability of various "how to" design methododologies such as Jackson, Yourden or Meyers [Friedman89]. It is interesting, however, that structured analysis techniques (eg. requirements modelling, requirements definition languages), though logically prior to other structured methods, only began to appear in the 1980s. These techniques, which aim to delimit the uncertainties of eliciting and negotiating users' requirements, are far less well established in commercial computing environments. Indeed, Friedman suggests that in the current "third phase" of computing - where meeting users' requirements has replaced the software crisis as the dominant constraint on progress - structured methodologies are being "dropped, or softened" in favour of prototyping techniques that enable more concrete communication with users in the design process [Friedman89 136].

Though the structured methodology framework continues to prove its value in the management of commercial computing, it has been adopted by practitioners quite selectively. For example, functional decomposition quickly became popular in the 1970s, because the procedure allows for many ways of splitting a particular process into functions [Bergland81] whereas structured analysis techniques are more talked about than practised [Friedman89]; modular programming was quickly adopted because it reduced the amount of change during maintenance [Friedman89] whereas Dijkstra's GOTO letter provoked more controversy amongst software scientists than practitioners [Pelaez88].

The "scientific" reform of programming practice has undoubtedly radically altered the management of software production. However as it gradually emerged, this framework assumed certain organisational forms for its social application in the workplace. Not all of these organisational assumptions have proved workable, or even fully relevant, in many software development installations [Friedman89].

2.3 Managing the software labour process

The introduction of structured methods did more than provide a scientific framework for project management. It also provided the opportunity for re-structuring the social division of labour. During the 1970s particularly, the pattern of re-structuring led theorists to analyse the labour process as one which was being progressively deskilled as a result of "scientific management", or Taylorist, techniques [Braverman74]. Since structured programming served to fragment programming into discrete tasks of greater and lesser complexity, managers now had the opportunity to allocate out particular tasks to particular programmers, according to criteria largely set by them. Throughout the late 1960s and 1970s, these criteria were, in general, informed by a desire to rationalise the skills of software staff and the flow of information between them. The aims were to increase general productivity by means of direct monitoring of programmers' output, concentrate the exchange of information amongst senior technical staff; and reduce dependence on any one programmer [Kraft77, Greenbaum79, Friedman89]. So, for example, in the Chief Programmer Team experiment, the effect was to generate a small cadre of senior technical staff who were both organisationally powerful and skilled in the management of structured methods, and a much larger body of junior, and less skilled, programmers who executed the tasks given to them [Baker and Mills73].

Though the deskilling struggles of the '70s are not a topic for this paper, it is worth recalling two key issues they raise in relation to the emergence of project management frameworks. First, deskilling and reskilling analyses highlight issues of organisational conflict: between programmers and their managers; between technical staff and users; between the computing function and senior management. Managing organisational conflict is undoubtedly a major preoccupation for industrial software managers. Since the science tradition rarely acknowledges the *power* relations of software production - most innovations with respect to the management of the labour process, or its organisational environment, have emerged as a outcome of industrial experience, rather than science-based reform. So, for example, managerial strategies to bypass the power of software staff in conditions of skill shortage, such as retraining clerical staff as in-house programmers or sub-contracting the work, are distinctly organisational solutions.

Second, labour process analyses highlight why and how the *criteria* for managerial strategies change, with changing organisational circumstances. If it was the case, during the 1970s, that attention was largely directed at fragmenting and re-structuring the labour process so as to control it more directly (eg. by instituting hierarchical chains of command, or by monitoring programmers' output according to certain metrics such as worksheets or lines of code per week) the pressures to intervene in this particular manner have since changed. More recent studies of the IT software labour process note that managerial strategies are currently geared to promote information flow, especially between design staff and users; to reskill development staff in application-specific know-how; and to increase the clout of IS Depts within the organisation [Child84, Lucas84, Friedman and Greenbaum84, Newman and Rosenberg85, Friedman89, Tierney91]. The key point here is that managerial intervention in the labour process is a process of continual - albeit partial - adjustment to changing technical, organisational and economic circumstances.

It is important to stress, however, that such interventions in the labour process are cumulative and incremental. New project management frameworks do not emerge from industry every few years, precisely because strategies - once implemented - tend to solidify particular divisions of technical labour or technical knowledge. Project management leaves marks on the organisation of software labour, and so constrains the extent to which non-incremental change is possible. Thus, even in rapidly changing organisational circumstances, the *framework* for project management

51

remains relatively stable. At this point in commercial computing history, we can identify a number of the common components of the framework, around each of which project managers can make a (limited) range of new adjustments:

- managing the division of labour (whether in teams, pools, matrix structures etc)

- managing project work with respect to the division of labour (dedicated project teams, "floating programmers" etc)

- managing hierarchies of command (within the computing function, and between computing and departmental users or external clients)

- managing the dissolution or the creation of sub-occupational specialisms, and the relationships between them (applications or systems programming, systems analysis, newer kinds of hybrid technical jobs such as business analysis. etc)

- managing the level, and kind, of reviewing to check the quality or quantity of work produced

- managing the balance of internal development work with external contracting or with the purchase of packages (assessing which work must be bespoke and which can be more economically bought)

- managing the relationship between a project's development, and the user it is intended for (specifying the problem, establishing working parties or liaison staff to elicit user requirements and/or to commit the user to the project, writing contracts to determine who will bear the costs of which part(s) of the project's development or maintenance, setting up mechanisms for user feedback, etc)

- managing the project's budget (using accountancy or other techniques to justify the project and to place estimates on its likely cost and duration etc).

2.4 The pursuit of adequacy in project management

Managers make decisions on the handling of each component in the light of their ability to deliver the goods, in the face of user demands, using whatever skills, organisational structures and techniques they currently possess. Project management is, in practice, a continuous balancing act where trade-offs are made so as to preserve the survival of the project to its completion. It is in this sense that adequacy is the aim of all project management, regardless of the application field. However, based on the issues raised in this section of the paper, it seems likely that the detailed *pursuit* of adequacy will differ quite profoundly between mainstream and high-integrity computing environments. And here, we return to one of the major implications of the distinction between the "scientific" and "industrial" management of software.

In mainstream computing, software scientists have provided the backdrop, rather than the plot, to the social shaping of the software development process. Industrial software management demands a view of the production process, not as the rigorous application of a fixed body of principles to well-defined and stable design specifications, but as a necessarily open-ended relationship with ill-defined user requirements and unstable organisational or market dynamics. The major priorities are to meet deadlines, control spending and, in so far as possible, answer user demands for working software [Friedman89, Pelaez88]. Given these priorities, the likely trade-offs in the pursuit of adequacy include relying on software staff who possess journeyman skills gained in commercial computing (as opposed to mathematical or engineering skills); doing some documentation (but not all); testing the software (but only to a certain extent); and implementing the software for use as soon as is practicable (even if that means considerable maintenance later). The kind of software produced in this kind of environment is likely to have had adequate quality controls imposed on it during the course of its development. But, by the same token, project managers in mainstream commercial computing are not likely to allow quality considerations to override the adequacy of the project as a whole.

In contrast, in high-integrity computing, it is quality which effectively defines the project manager's brief. So, although extremely rigorous design and testing techniques may get fairly short shrift in mainstream computing, there is a strong commercial incentive in high-integrity computing to arrange the labour process around the realisation of quality through

53

"scientific" means. In consequence, it is likely that at least some specialists employed on projects of this kind work in a labour process which is unfamiliar in mainstream computing, as we will see in the next section.

3 A sketch of "scientific" software development: formal methods, and its practitioners, in high-integrity computing

In the last section of the paper, I suggested that the history of project management strategies has been shaped by both "scientific" and "industrial" solutions to software production problems, but that for the most part, the latter have come to dominate mainstream practice. Most of the current array of software managerial practices are direct responses to commercial and organisational pressures on the software development process; by comparison, the scientific control of software production has proved to be a weaker stimulus [Pelaez88].

In this section I want to suggest that in high-integrity computing projects, it is the *scientific* approach which promises practical answers to the commercial need to demonstrate how reliability requirements are being met, and this alters the development environment to which safety-critical managers must respond. We find that distinctively "scientific" work characterises at least some pockets of safety-critical development activity - formal methods, and its practitioners, being a case in point.[2]

Though formal methodists constitute only a part of the skilled work-force deployed in high-integrity computing, and formal methods are only one in an armoury of techniques used to increase confidence in the trustworthiness of a project's output, it is useful, for two reasons, to single them out. First, high-integrity computing is currently the most important market for formal methods, and formal methods are becoming a major contributor to satisfying the demand for trustworthiness common to all safety-critical projects [Jackson89, Bloomfield et al.91, MoD91].[3] Secondly, however, formal methods are often regarded with scepticism by the computing

---

[2] The notes offered here do not distinguish between different kinds of formal method. There is a world of difference between, say, formal specification and automated theorem-proving, though both fall under the general rubric of formal methods. In addition, formal methods are applied to different components of a system which introduces important variations in their purpose and use. Again, these are not explored here.

[3] Over the last two years at Edinburgh, Donald MacKenzie, Eloina Pelaez and myself have been investigating the genesis and use of formal methods, under the ESRC-funded Programme on Information and Communication Technologies (PICT). This research, which will continue until 1994, is based on an exploration of detailed case-studies of formal methods R&D and implementation, many of which lie in safety-critical fields.

54

mainstream. They are seen as running counter to normal practice, with respect to the organisation of development, the skills required, and the costing and scheduling of projects [Coleman90, Hall90, Harding and Gilbert91, Tierney92]. Thus, a crude profile of formal methodists in high-integrity projects - in terms of the projects they are engaged in, the clients they serve, the kinds of expertise they deploy and the information networks they enter - serves to contrast (part of) the socio-economic constituency of high-integrity computing with that found in the computing mainstream.[4]

3.1 The use of formal methods in high-integrity projects

Any high-integrity computing project encounters two formidable difficulties. First, the field is new. Though safety engineering is well-established - ancient, even - the organisational and sectoral pressures which make it attractive to locate control over crucial elements of a system within a computer are modern phenomena. On this account, a high-integrity project is one which treads new ground. Its actors must write the rules and build the tools as they go along. Second, despite the extreme uncertainties in doing this work adequately, the price of failure is high. Thus, everything from the conceptualisation of the specification model through, in some cases, to the proving of the underlying algorithms must be understood, built, tested, reviewed, rebuilt and found fit.

Not surprisingly, the net effect of these two characteristics is that most high-integrity computing projects are path-breaking in their conception and execution. The formal methods used within them are embedded in projects which carry all the additional costs associated with uniquely difficult developments.

Though we can assume that, given time, some of the costs associated with projects of this kind will be reduced - formal models and tools (or bits of them) are already being reused - that is not likely to happen quickly or uniformly. The clients who fund formal techniques and tools for use within high-integrity projects do so in terms of meeting their own immediate specialised needs. That is, unless the formal methods work is being applied to something, it will not be funded.

---

[4] For purposes of brevity, this sketch does not distinguish between different kinds of high-integrity project, though I recognise that not only does the context of high-integrity computing vary enormously across sectors and applications fields, but decisions as to where to locate the bulk of a system's reliability requirements (eg. in its software, or elsewhere) also vary.

One effect of this is that each project using formal techniques yields a model or language or tool specially geared to its particular purpose. Any additional work on increasing its potential flexibility or general robustness is something which happens as a secondary and slower activity. Since anything except partial re-use (particularly across, rather than within, contracting companies) is frequently an option which is fraught with potentially serious design flaws, it often - ironically - becomes more cost-effective to engage in new development rather than to tailor others' work. In addition, many of the formal techniques and tools developed for clients in defence departments are subject to restrictions on circulation, use or sale elsewhere, especially across national boundaries. Thus, the path-breaking quality of many high-integrity projects is reproduced, rather than assuaged.

Given the typical sorts of requirements demanded by high-integrity funders and clients, formal methods, in this niche, are a set of techniques which are useful for restricting systems: they aid the prevention of the system doing something undesirable. While the clients certainly demand functionality from their systems, their interest in sponsoring formal methods R&D does not generally spring from this aspect of their requirements. Thus, some of the most noteworthy technical and procedural achievements of formal methods (eg. automated theorem-provers; system-centred strategies for dove-tailing formal analysis with testing; organisational routines for independent review of the designer's models and axioms) reflect the preventive focus demanded by these particular clients. Shaped by this history of use, formal approaches - with their attendant tools and work practices - most visibly prioritise assurance over functionality. They "stand for" the polar opposite of the enabling techniques sought in mainstream practice.[5] That is, for those who are not in the know, the bulk of evidence points to formal methods as techniques whose primary value is restricting software from working wrongly, rather than enabling software to work adequately.

The cost-effectiveness of the formal methods component of high-integrity projects is adjudicated by the client on the basis of its contribution to meeting stringent design requirements. If it manages that, it has paid its way. Thus, the design of formally-based tools is not usually premised on criteria such as the tool's ease of use by (non-expert) others, or by the time or

---

[5] To those who use formal methods, this may be easily seen as a false dichotomy. Our research group frequently encounters persuasive arguments that what a formal approach offers is clarity. This clarity of conception, most especially through formal modelling of a specification, can, in principle, be as easily applied to capturing functional requirements as to restricting functionality for purposes of safety or security.

space it needs to complete its tasks. For them to work efficiently, the current state of many verification tools demands that the manipulators of the tools are themselves experts and/or that the amount of memory space available for the tool to work in, is not a critical factor.

3.2 The clients for formal methods components of high-integrity computing projects

Without the funding which has been forthcoming from large government departments (most notably, defence) over the last 20 years, it is most unlikely that research into formal methodologies, languages and tools could ever have evolved beyond a few isolated projects; academic not merely in location, but also in scale and exploitability. In the case of formal methods, agencies attached to the military (eg. RSRE, CESG, CSE, NSA and its offshoots) have been crucial to their evolution as exploitable techniques. These, together with central government agencies for other sectors (eg. NASA, CAA), constitute the vast majority of the clientele for projects which are premised on the use of formal methods.

As customers they are somewhat untypical for they are sponsors, as much as purchasers, of formal methods. Their requirements focus on problems of security or safety: cost control (though always desirable) is not the main goal. The trustworthiness of a computer-based system is an attribute which is everywhere inscribed in the process by which that system is built and checked. Thus, what these clients contract for is detailed documentary evidence of that process, as much as for any final working product. For these clients, it is not sufficient for the product to behave, it must be known to behave.

Thus, the relationship between themselves and their contractors is more likely to be collaborative than wholly transaction-based; developmental, rather than black-boxed, artefacts are acceptable as output. Finally, having a role as regulators, as much as procurers of high-integrity systems, these clients can initiate certification standards which solidify particular "visions" of formal methods within particular fields - for example, the Orange Book with its approved security models and tools; 00-55 with its detailed formalisation of the management of safety [DoD85, MoD91].

3.3 Formal methods practitioners

The profiles of those engaged in the design and implementation of formal methodologies and tools depart significantly from the mainstream of software developers. Most usually, they are employed in "laboratory" companies, or in dedicated departments within large organisations, which enjoy close associations with universities, and whose purpose is to conduct innovative R&D which may (or may not) result in commercial products. In this sense, the developers are applied academics, whose design skills are firmly grounded in the theoretical backgrounds from which their interest in formal methods has evolved (eg. mathematics or computer science).

As with academic scientists, many hold doctorates and their career routes have been structured along highly specialist pathways. Their expertise is one of mastery of formidable bodies of theory and its application, acquired over a long period of formal education. As with any specialism, these developers are practised in their own occupational argot: the terms and symbols and concepts, so troubling and arcane to outsiders, are the necessary disciplinary short-hand through which formal developers progress their ideas with each other.

Like academic scientists also, the organisation and monitoring of formal developers' work is, most typically, peer-oriented. The audience and judge of their work is less their immediate management than their occupational peers (who may be their clients, collaborators or competitors) to whom they routinely circulate a variety of written and verbal reports on their progress. This strategy of *independent* review of work which is so characteristic of formal methods work in high-integrity computing - yet is so alien to mainstream development practice - trades directly upon an academic rather than an industrial kind of labour process.

3.4 Their information networks

Given the general characteristics of the key actors, outlined above, and the sorts of specialised design problems upon which they work, it is not surprising that the work-related networks of formal methodists are unusually tightly bounded. It is an occupational community which speaks mainly to itself or to other non-formalist engineers engaged in safety- or security-critical work. They attend the same conferences and workshops; subscribe and submit papers to the same journals; peer review each other's work; visit each other's work sites. While some pockets of this community -

notably, those engaged in classified security-critical defence projects - do not discuss their progress in using formal techniques, the whole field of formal methods R&D gains substantially from the tightness of the network.

What can be efficiently traded through a network of peers, who know each other or of each other and who hold certain basic problems in common, is *detail*. Indeed, it is knowing the detail which guarantees membership as an insider: those furthest from a field of scientific endeavour will not be aware of the specific potentials and uncertainties which the field's "core set" are privy to [Collins85, MacKenzie90].

The flip side, of course, is that a tight network is exclusive: learning from the detail assumes that the members of the network engage in the same sorts of problems, speak the same language, share the same basic assumptions. This pattern of information flow is, again, much more akin to academia than to industry. User managers, project managers and programming staff engaged in mainstream software development cannot easily participate, even if they wished to. And, thus, the social boundaries which define the formal methods community also serve to insulate that community from the main computing fraternity.

4 Some research questions on the management of safety-critical computing projects

The distinctive characteristics of formal methodists, outlined above, suggest a social process of software production (and user consumption) in high-integrity computing which differs significantly from normal industrial environments [Friedman89]. We can assume that formal methods, amongst other checking and testing techniques, have been able to grow commercial roots in this niche (far more so than in other application fields) precisely because of their value in addressing reliability requirements.[6] Thus, unlike many mainstream application environments, there is a need in safety-critical projects for this archetypically "scientific" production process, so long as it contributes substantially to the production of safer software. In this niche, the requirement for trustworthiness stands equal to the cost, time and user constraints which have shaped mainstream computing.

---

[6] This is not to suggest that mainstream computing is not also concerned with the quality of its software. However, definitions of software adequacy in high-integrity computing will be different - and necessarily higher - than definitions of software adequacy found in sectors or applications fields where the costs of error are not as large.

If the development environment of formal methodists is any guide, we can expect that managing safety-critical projects involves mobilising and controlling unusual clusters of specialists whose working practices differ substantially from those of industrial programmers, analysts or systems engineers [Tierney91]. In addition, safety-critical managers must cost out and co-ordinate a development process whose commercial dynamics - especially with respect to the commissioning and delivery of safety-critical products - differ markedly from normal trading practice [Brady et al.92]. Finally, safety-critical managers must grapple with the novelty of safety-critical computing. As we saw in section 3.1, most high-integrity projects are path-breaking in their conception and execution; each one cannot readily be compared to anything else. Possessing unique characteristics, these projects are likely to remain expensive and experimental, since each one reproduces, rather than assuages, the need for specially-geared tools and techniques.

In consequence, we can hypothesise that when safety-critical project managers engage in the staples of good project management - budgeting and scheduling the work, co-ordinating the labour process, evaluating the project's progress, etc. - they must do so to fit contexts where:

- conventional accountancy criteria are secondary to the search for greater assurance;

- specialist, rather than generalist, technical skills are high;

- links to academia are much closer than is the norm in mainstream system development;

- the documentation of the design process is as important to clients as the product itself;

- developmental rather than black-boxed artefacts are acceptable as output.

These contingent factors - which imply unusual forms of work organisation and technical practice - seem likely to make it especially difficult for safety-critical project managers to simply emulate the managerial strategies which have evolved in mainstream computing. For example, it may prove difficult to show, by means of standard cost-benefit or scheduling techniques, that the work is being efficiently managed, or that its costs are justified. Likewise, certain common managerial choices made over the division of labour (eg. routinising sections of the work, and devolving it onto inexperienced design staff) may, in projects of this kind, prove counter-productive, dangerous and an underuse of (expensive) skilled labour.

Clearly, the management of safety-critical projects carries special difficulties. The rules of best practice in this field are still evolving, and the price of failure is high. Neither the development nor the evaluation of safety-critical systems is simply a technical matter. The best development tools are worthless if those using them do not have the appropriate skills or if organisational procedures are inadequate. In addition, it is not enough for safety-critical systems to be trustworthy. They have to be *seen* (by clients, regulators, and sometimes the general public) to be trustworthy. This inevitably involves different perceptions of acceptable risk [Douglas86] and different ways of demonstrating trustworthiness, using an array of testing and verification techniques.

At Edinburgh University, we have recently embarked on a three-year sociological study of the different development and evaluation environments which pertain across a number of sectors deploying safety- or security-critical computing applications. In the course of this project, we hope to analyse the social processes shaping the development and evaluation of high-integrity computer systems - including an investigation of the managerial procedures invoked to co-ordinate and control the emergence of new systems into working use. The result of the research proposed will be a "map" of how technical issues intertwine with commercial and organisational concerns, across a number of sectors, which may prove to be of interest to managers and practitioners who are currently seeking to define best practice for their own sector.

References

[Baker and Mills73]   Baker FT and Mills HD: "Chief Programmer Teams." in Datamation, December, pp.58-61, 1973

[Bergland81]   Bergland GD: "A Guided Tour of Program Design Methodologies." in Computer, October, pp.13-37, 1981

[Bloomfield et al.91]   Bloomfield R, Froome P, and Monahan B: "Formal Methods in the Production and Assessment of Safety-Critical Systems" in Reliability Engineering and Systems Safety Vol 31 No 1-2 pp. 51-66, 1991

[Brady et al.92]   Brady T, Tierney M and Williams R: "The Commodification of Industry-Application Software." in Industrial and Corporate Change, Winter, 1992

[Braverman74]   Braverman H: Labor and Monopoly Capital. Monthly Review Press, New York, 1974

[Brooks75]   Brooks FP: The Mythical Man-Month. Addison-Wesley, Reading, Mass., 1975

[Brooks86]   Brooks FP: "No Silver Bullet - Essence and Accidents of Software Engineering." in H-J Kugler (ed) Information Processing 86. Elservier Science Publishers (North Holland), 1986

[Buxton90]   Buxton JN: "Software Engineering - 20 Years On and 20 Years Back." in Journal of Systems Software Vol 13 pp. 153-155, 1990

[Child84]   Child J: "New Technology and Developments in Management Organisation." in OMEGA: International Journal of Management Science. Vol 12 No 3, 1984

[Coleman90]   Coleman D: "The Technology Transfer of Formal Methods: What's Going Wrong?" paper presented to the Workshop on Industrial Use of Formal Methods, Nice, March, 1990

[Collins85]   Collins H: Changing Order: Replication and Induction in Scientific Practice. Sage, 1985

[Dahl et al.72]   Dahl O-J, Dijkstra EW and Hoare CAR: Structured Programming. Academic Press, London, 1972

[Dijkstra69]   Dijkstra EW: "On the Interplay between Mathematics and Programming." in Program Construction, Lecture Notes in Computer Science. Springer, New York pp.35-46, 1969

[DoD85]   Department of Defense: Trusted Computer System Evaluation Criteria. Department of Defense, December, 1985

[Douglas86]   Douglas M: Risk Acceptability according to the Social Sciences. Routledge, 1986

[Fleck and Tierney91]   Fleck J and Tierney M: "The Management of Expertise: Knowledge, Power and the Economics of Expert Labour." Edinburgh PICT Working Paper, No. 29, 1991

[Friedman89]   Friedman A: Computer Systems Development: History, Organisation and Implementation. John Wiley & Sons, 1989

[Friedman and Greenbaum84]   Friedman A and Greenbaum J: "Wanted: Renaissance People." in Datamation. September, 1984

[Greenbaum79]   Greenbaum J: In the Name of Efficiency. Temple, Philadelphia, 1979

[Hall90]   Hall A: "Seven Myths of Formal Methods." in IEEE Software, September, 1990

[Harding and Gilbert91]   Harding S and Gilbert N: "Taking Up Formal Methods." paper presented to the SPRU/CICT Workshop on Policy Issues in Systems and Software Development, July, Science Policy Research Unit, Brighton, 1991

[Ince and Andrews90]   Ince D and Andrews D (eds): The Software Life-Cycle. Butterworths, 1990

[Jackson89]   Jackson M: "Formal Methods and Critical Software Development" in SafetyNet 89 Conference Proceedings, RSRE, on Industrial Experience of Formal Methods. SafetyNet, Worcester, 1989

[Kraft77]   Kraft P: Programmers and Managers: The Routinization of Computer Programming in the United States. Springer, New York, 1977

[Lucas84]   Lucas H: "Organisational Power and the Information Services Department." in Communications of the ACM. Vol 27 No 1, January, 1984

[MacKenzie90]   MacKenzie D: Inventing Accuracy: A Historical Sociology of Nuclear Missile Guidance. MIT, 1990

[Macro and Buxton87]   Macro A and Buxton J: The Craft of Software Engineering. Addison-Wesley, 1987

[MoD91]   Ministry of Defence: "Interim Defence Standard 00-55 - The Procurement of Safety-Critical Software in Defence Equipment." MOD, April 1991

[Naur and Randell69]   Naur P and Randell B (eds): Software Engineering. Report on a conference sponsored by the NATO Science Committee, Garmisch, Germany, 7-11 Oct 1968. Scientific Affairs Division, NAT0, Brussels, 1969

[Newman and Rosenberg85]   Newman M and Rosenberg D: "Systems Analysts and the Politics of Organisational Control." in N Piercy (ed) The Management Implications of New Technology. Croom Helm, 1985

[Pelaez88]   Pelaez E: A Gift from Pandora's Box: the Software Crisis. PhD Thesis, Edinburgh University, 1988

[Sommerville89]   Sommerville I: Software Engineering. 3rd Edition. Addison-Wesley, 1989

[Tierney91]   Tierney M: "The Formation and Fragmentation of Computing as an Occupation: A Review of Shifting Technical Expertise." Edinburgh PICT Working Paper, No. 25. Edinburgh University, 1991

[Tierney92]   Tierney M: "Formal Methods of Software Development: Painted into the Corner of High-Integrity Computing?" in Ryan P and Sennett C (eds) FM91 Workshop Proceedings, Springer, 1992

PART II

CURRENT RESEARCH

Wednesday 10th February 1993

Classification
of
Programmable Electronic Systems Operation
for
Testability

Dr. J.S.Parkinson C.Eng. MIMechE

NEI-Control Systems Limited

Rolls-Royce Industrial Power Group

Abstract

This paper considers the requirements for effective testing of Programmable Electronic Systems. It discusses how interfaces available to the test environment may vary with system development and how these interfaces together with the mode of operation of Progammable Electronic Systems affect the ability to perform dynamic tests. A number of desirable test attributes are defined and four modes of operation are compared with respect to these attributes.

1   Introduction

The aim of this paper is to highlight and discuss the modes of operation of Programmable Electronic Systems with respect to testability, particularly with respect to controllability and observability of the dynamic test process.

Let us remind ourselves of what we aim to do during testing. We aim to demonstrate compliance with a specification. In order to do this the specification will have been interpreted to produce test cases. Test cases will have pre and post conditions. Thus we must establish and observe the pre conditions and observe the post conditions. We must know when the pre conditions are met so that we can start execution of the process and we must know when the post conditions should have been met so that we know when to record the test results. Establishing pre conditions during testing will depend on the controllability of the system under test, which will in turn depend on interfaces provided and on the logic implemented within the PES.

It is not the aim of this paper to discuss the construction of testable logic, details may be found elsewhere [IEE88, Lew85], nor is it the aim of this paper to establish the desirability of PES modes of operation with respect to plant performance or integrity, however it is necessary to discuss the PES mode of operation in the context of system function. The nature of the system function influences the test cases required, the test cases required may induce different behaviour in different types of PES.

One approach [MOD91] to testing a system beyond its positive requirements is to use random inputs with the system in operation. Both the timing and levels of signals may be random. With such testing there is the possibility to exercise the system under test in an unrealistic manner, in that combinations of plant signals being randomly distributed do not fall within a "typical" operational envelope. This may be seen as an advantage: the system is being tested outside its normal operational envelope and the sensitivity to signal corruption investigated: an alternative view is that the distribution of the plant signals may be a physical impossibility and that the test has no useful interpretation. This paper will discuss how the mode of PES operation may determine the functional behaviour of the system when random inputs testing is applied.

Regardless of how signals are injected we must be able to observe the response of the system and determine the stimuli which initiated the change in outputs. The effect on these activities by the PES's mode of operation will be considered.

Systems developed in a modular fashion allow for module and sub-sequently sub-system and system testing. This paper considers aspects of PES operation which influence the ability to perform testing in this progressive manner. The concept of a moving test boundary is employed as a framework for discussing PES interfaces which affect construction of the test environment.

There are similarities between the MASCOT [All87] and the Real Time Structured Analysis (RTSA) design methods. Both methods use the concepts of processes, channels (data-flows) and pools (data-stores). In MASCOT, process synchronisation is governed by WAIT and SIGNAL events. When a process encounters a WAIT event it suspends activity until a corresponding SIGNAL event occurs. The SIGNAL event may be triggered by other processes or by external events (interrupts). A similar interpretation may be placed on control flows in RTSA. These concepts are useful in discussing the internal behaviour of the PES and will be used to "model" and classify PES modes of operation. In this paper RTSA conventions for control and data flow models are used to describe the behaviour of PES "operating systems". The aim of generating these simple models will be to identify how, in broad terms, the execution method of the PES affects our ability to perform tests.

The paper starts with a discussion of what is required to observe the execution of a test. In Section 3 interfaces available during the evolution from module to integrated system are identified. In the light of these discussions certain testability attributes are identified, the modes of operation of a number of types of PES are considered with respect to these attributes, and a primitive scoring system is employed in an effort to compare the PES types.

2   Monitoring Tests

Figure 1 illustrates the relationship of a monitor with elements of a PES to be tested. For a description of the mode of operation of the PES see Section 4.

The monitor process must be able to observe the input and output buffers and the system memory. It must also be able to export the observed data to some external equipment, labelled in Figure 1 as analysis equipment. The testability of the system is very much determined by the performance of the monitor and its interface to the external analysis equipment. The speed at which the monitor can execute determines the resolution of observations on the system under test. There are two modes of operation for the monitor process, it may operate as a sampler or it may run as an event trapper.

As a sampler the monitor must scan input and output buffers and memory and transmit data to the analysis equipment at predetermined intervals. Signal Processing theory dictates that sampled observations must be made at, at least, twice the frequency of the highest frequency signal of interest. This would suggest that the monitor process must run twice for every execution of the fastest process in the system, and furthermore it must be interleaved with the execution of the fastest process. To simplify analysis the monitor must be run at fixed intervals. These requirements are probably more onerous than those for the system under test. Implementation of such a monitor embedded within the target system is undesirable as it will either affect the system performance or require capacity, for test purposes, well in excess of that required by the application. An alternative would be to design the system so that the data stores indicated are observable by an external monitor. Dual ported memory would be one way of allowing access to data at a suitably rapid rate.

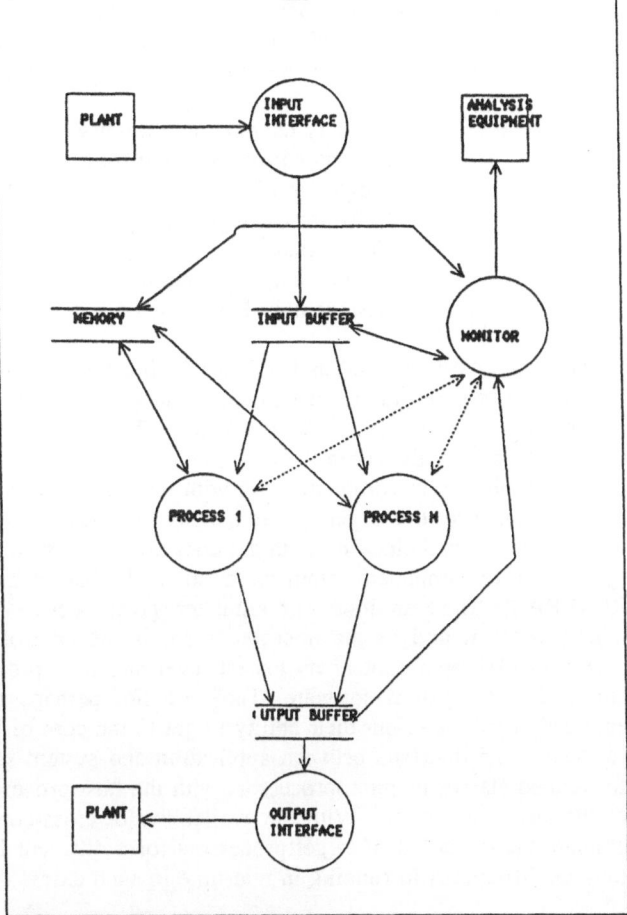

Figure 1 : A Monitor Process

As an event trapper the monitor must observe the input and output buffers and memories very quickly in relation to execution of code. It compares the state of the data stores with a state of interest, if it detects the state it transmits the data to the analysis equipment. In large systems it may not be possible for the monitor to scan and mask the system data stores before they change. In order to reduce the amount of data to be scanned by the monitor it may be possible to select only a subset of data stores which are regarded as critical to the observation of an event.

It is clear that there is some need for synchronisation between the monitor process and the process(es) under test to allow unambiguous results to be recorded. There must also be careful consideration of signal, process and monitor dynamics, so that one can set realistic targets on each. Ideally the monitor process would be able to control execution of the processes under test. Control of execution by the monitor is desirable for a number of reasons: firstly it would aid integration testing as processes could be loaded and "switched-in" as each process is tested in conjunction with others. One may want to place breakpoints in the process under test to allow the monitor to observe intermediate values. And one may want to advance/retard one process with respect to others (in an asynchronously operating system) in order to investigate the effect.

In-circuit emulators provide many of the features of a monitor process as described above. Emulators usually have a monitor program which provide the following:

- Memory examine and modify
- Emulator Processor register examine and modify
- Single Step Execution
- Software Breakpoints
- Trace Analysis
- Logic Analyzer Functions
- Disassembly Functions

They may be employed as test-bed for embedded microprocessor systems. Use of an emulator provides the facility to replace target ROM and RAM with host RAM. The replacement may be selected by address. This allows the emulator to simulate the actions of peripherals within the target by manipulating the contents of addresses used by the peripherals in communicating with the microprocessor.

Emulators use the target system's clock to control execution in real time or it may use its own internal clock if the target does not yet exist, in fact the whole of the target system can be simulated within the emulator by use of host RAM in place of target ROM/RAM. Thus an in-circuit emulator gives access to the core of a PES, the microprocessor, and as the microprocessor is the controller of the PES it provides access to PES operation at its lowest level and thus provides low level diagnostics appropriate to system software. They are not perhaps the easiest way of testing application software, but their ability to get to the core of a system make them useful for testing the interface between application and system software. They are available for related classes of microprocessors with the host processor being in the same class as the target processor. With the availability of cross-compilers host processors can emulate the operation of target processors for a different family of processors. There may be difficulties in running in real-time in such cases.

3   Test Boundaries during System Integration

Consider Figure 2, the boundaries of the system at module, sub-system and system test phases are illustrated. The items drawn within the boundaries constitute the system under test, the items without form the environment of the items under test. In some cases these items are not available or amenable for use as drivers for executing the test. In such cases a replacement will be required, this replacement can be regarded as a simulation of the environment. Clearly the extent of the simulation will change as the system boundary changes, however the means by which the simulation may inject signals may remain the same if the physical representation of the control and data flows crossed by the system boundary remain unchanged. At the system testing phase it is likely that the simulation will drive input ports which will be used in system operation.

3.1   Module Test Boundary

In order to identify the requirements for test environment construction it is necessary to inspect the data and control flows which cross the test boundaries. In Figure 2 the test boundary crosses the data flow from the input interface to the input buffer. This

70

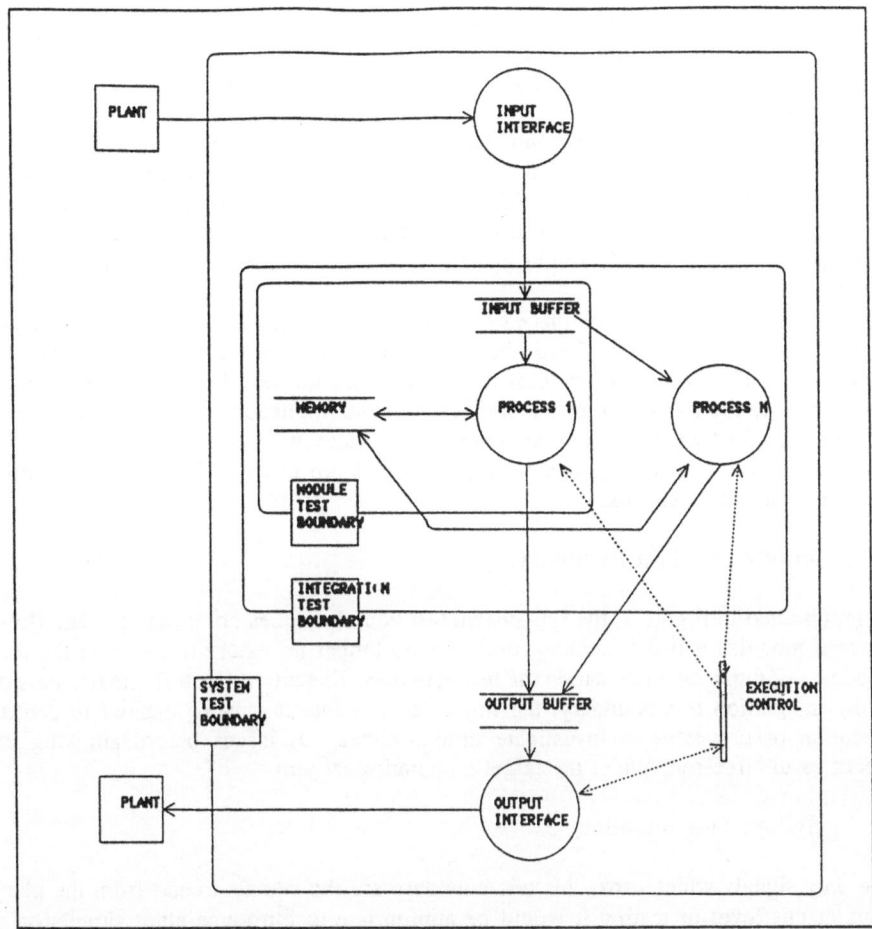

Figure 2 : Evolution of the Test Boundary

implies that the test environment must write directly to the input buffer. As the input buffer will be an allocated area of memory the test environment will most easily be realised by software.

By enclosing the system memory within the test boundary it is implied that the item under test resides in the target system. (An inner boundary around just the module under test would imply that the target system memory could be simulated as would be done during static analysis. Use of an in-circuit emulator would also allow simulation of target memory).

The item under test may write data to the memory and so it must be read by the monitor process as indicated in Figure 1. The simulation of signals from other parts of the software system must be generated, as indicated by the data flow which crosses the test boundary between Process n and the memory . The implication of Figure 2 is that all processes communicate via a global database. If a dataflow existed between the module under test and other modules then the implication is that data is passed directly as procedural parameters. This may be effected either by a CALL to or from the module under test. If the dataflows are provided by CALLs from the module under test then there is a need for stubs to simulate the responses of the called modules. It is possible that communicating processes are implemented on separate processors, as in a Distributed Control System (DCS) for example, in this case a PES-PES interface would cross the test boundary.

Control flow from the Execution Control bar also crosses the test boundary, clearly, the test system must be able to initiate execution of the module under test, but it must also be able to reproduce WAIT/SIGNAL controls as required by the module under test, this could include both application control signals (e.g. signal from a plant event recognised by other processes) and signals to represent scheduling of the task within the operating system.

The output buffer has been placed outside the system test boundary. This does not imply that the output buffer need be simulated, it will be used as part of the test environment. The output buffer must be inspected during test, however as there is no flow back into the process from the output buffer the implication is that any reading of the output buffer by the test environment will not affect the operation of the item under test; there is no need to reproduce the dataflows from other processes to the output buffer for the same reasons.

3.2 Sub-System Test Boundary

As represented in Figure 2, the sub-system test boundary does not cross the data flows between modules and the memory, thus it is no longer necessary to simulate the data responses of those modules within the test boundary. Execution Control remains outside of the integration test boundary, the implication is that it may be desired to control execution of processes to investigate timing/rendezvous issues before allowing the processes to "free-run" under the target's operating system.

3.3 System Test Boundary

The only signals which cross this test boundary are the signals to and from the plant. Thus at this level of testing it would be appropriate to introduce plant simulation to generate consistent sets of signals. However, simple switch and lamp test rigs could also be used to drive the PES-Plant interfaces.

4 PES classification by operation

In order to compare types of PESs with respect to testability a number of desirable attributes have been identified. As the paper is concerned particularly with the mode of operation of PES's it is the sensitivity of each attributes to time related issues that will be considered.It is assumed that the process logic will be the same regardless of the system upon which it is to be implemented. Thus the effect of the process logic upon controllability and observability, which may be profound is not considered here. It is worth noting that this is a comparative study, where an advantage of one over others is identified it will be noted.

Controllability
We need to consider whether the mode of operation aids or hinders the ability of the tester to inject signals which drive the system to the desired state.

Observability

We need to consider whether the PES's mode of operation makes it harder for the tester to record the signals required to provide evidence that the system is in a desired state.

Traceability

We need to consider whether the way in which the PES operates makes it difficult to record the signals necessary to demonstrate that a particular path has been executed. We must assume that adequate "instrumentation" has been included in the process logic to ensure that unique evidence is available in the data areas available to the monitor.

Predictability

We need to consider whether the PES's mode of operation provides for a means indicating when execution of logic is complete, so that we know when to observe the outputs.

Repeatability

We need to consider whether the PES's mode of operation is sensitive to slight variations in timing. Remember we are not considering whether it is desirable from the point of view of the application for the functional behaviour of the system to vary with slight variations in timing, we are considering how sensitivity to timing of signals makes life difficult for the tester.

Functional Testability

We must consider how the mode of operation affects our ability to inject and record the signals associated with the input to output transformation of a functional "block".

Accessibility

We need to consider the ease with which signals from a test environment may be introduced to the items under test.

Let us now consider the operation of four types of PES.

4.1   Synchronous Operation Single Tasking (Type A)

4.1.1   Description of behaviour

Figure 3 represents the operation of a synchronous system. Each of the processes indicated on the diagram operate in sequence. The input interface is signalled to start a scan of inputs, placing the data captured in an input buffer. Once all inputs have been scanned, the control process is signalled. The input interface then waits for a signal to restart. The control process then signals the application process to commence. The application process operates on the data in the input buffer and any available data held in memory. As each step in the process completes, data is written to the output buffer. Once the application process reaches the terminating statement, it signals to the control process that it has completed. It then waits for a signal to restart. The control process then signals to the output interface to commence operation. The output interface takes data in the output buffer and converts this data into a form which can drive the plant. Once the output interface has processed all data in the output buffer it signals to the control process that it has completed. The control process then usually waits for a signal to restart. This signal is usually triggered following an elapsed time. This type of operation is typical of Programmable Logic Controllers and Digital Process Controllers.

4.1.2 Discussion of Test Attributes

Controllability
Given that each system must accept input signals and that in driving the system to a desired state we are permitted to have strict control of these signals, this mode of operation is not perceived to offer any particular advantages over type B, C or D systems.

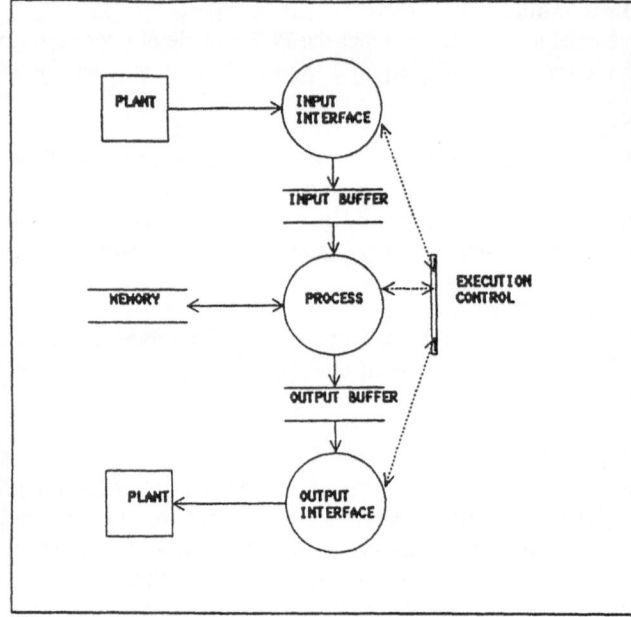

Figure 3 : Synchronous Operation

Observability
Outputs change in a short well defined time span, this will make it easier to record outputs as they will be steady for a known period of time, the duration of the process execution time. The observability of the process logic is not considered here as it would be a coding/design issue common to each type of PES.

Traceability
Recording inputs and outputs and memory contents at the start and end of the process execution will allow determination of the execution path (assuming adequate "instrumentation" of the process logic).

Predictability
The process under this mode of operation will be executed at regular intervals. This will help in knowing when to record the contents of buffers and memory.

Repeatability
The sampling and latching of the input interface at predetermined intervals desensitizes the process logic with respect to timing of input signals, provided they are applied at any time before the scan of inputs, the process logic will operate in the same way.

Functional Testability
In order to ascertain that the system has produced the correct transformation of inputs to outputs it is necessary to record the inputs as seen by the process logic. As the inputs are all latched at the start of execution the monitor is free to record them at any time

before the next scan of the inputs in the knowledge that the signals recorded are those employed by the process logic at all stages of execution. This means that it is not necessary to time stamp the inputs. This is seen as an advantage over type B and C systems.

Accessibility

Because the system is single tasking, it is not possible to run a separate task within the system which can manipulate the system memory. This makes the system less accessible than type C and D systems. The system could be run in debug mode whereby each instruction is executed and control returned to the debugger. Within the debugger one generally has the ability to manipulate memory. If the scan time of the application program is well within the capabilities of the hardware platform it may be possible to embed within the application code calls to simulation subroutines which can manipulate system memory, these simulation subroutines would form part of the application software; they would have to be disabled in the delivered system. In order to maintain timing of the "as tested" system, delays would have to be introduced to replace the execution of the simulation procedures.

Rather than interleave the simulation with the application code, it would be possible to run the simulation procedures at the end of each scan, (again providing that there is sufficient spare processor time before the commencement of the next scan). The simulation procedure could manipulate memory using hard coded procedures: a more flexible approach would be for the embedded simulation routine to service a communications channel, interpreting incoming messages and manipulating memory accordingly. We should be very careful about adopting such an approach if the simulation is interleaved with the application process as it would in effect introduce asynchronous input to the system. This may be exactly what we wish to avoid, its avoidance being the primary reason for selecting such an operating system in the first place.

These methods may be unattractive in that they affect the application software, increasing its complexity and introducing the possibility of incorrect operation in the field (if the simulation procedures are erroneously switched in). To avoid such worries then the simulation must access the system via existing interfaces. The difficulties in introducing appropriate signals to various interfaces are discussed in [CON92].

4.2   Asynchronous Inputs in a Single-tasking system (Type B)

4.2.1   Description of behaviour

Figure 4 represents the operation of a system in which the contents of the input buffer may change during the execution of the application process. The mechanism which allows this to occur is not the concern of this discussion. In other respects the operation of the system is similar to that for the system represented in Figure 3.

In order to make such a system operationally identical to that for Figure 3, the simple measure of making local copies of the input buffer within the application process would suffice. Memory constraints may prevent this measure from being taken.

Consider a process which uses a plant input for the basis of a number decisions. Using a system such as that shown in Figure 4 introduces the possibility of parts of the process using different values for that input if the input updates during the execution of the process. This may result in the generation of a set of inconsistent control signals which persist until all sections of the code have been executed using the same value of input variable. This could be a considerable time if the process has moved to a state where the "inconsistent" branch is no longer executed. It should be clear that from the system security point of view such a method of operation is extremely undesirable.

Although it can never be guaranteed that the output changes are all made at the same instant, it is assumed that the execution of the output interface is very rapid when compared with

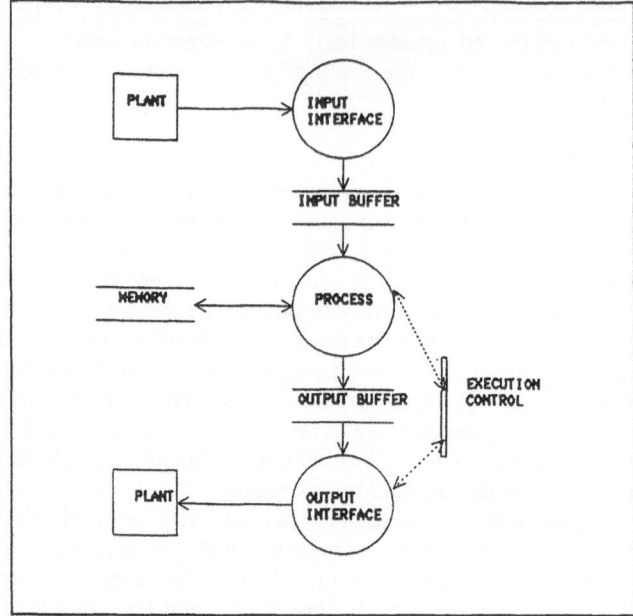

Figure 4 : Asynchronous Inputs

plant dynamics. This would ensure that the time taken for the PES outputs to update would be small thus reducing the time period during which there is a possibility of inconsistent control signals being applied. If however outputs were allowed to change at any stage during the process execution then the "window of opportunity" for output mismatch would depend on the execution time of the process.

Given these doubts about the propriety of such a method of operation, the aspects of testing such a system will be discussed.

4.2.2 Discussion of Test Attributes

Controllability
As for type A systems.

Observability
As for type A systems.

Traceability .
In this case time stamped recordings of input data will be required so that synchronisation of input data with process execution, which must also be recorded, can be achieved during analysis of the test results. For example if we are looking for evidence that a particular branch, which is dependent on a value of an input, was executed we would need to confirm that the input conditions for the branch were achieved; if they were not, due to the necessary signal being applied after the conditional statement being executed for example, it would be futile to search for the evidence.

76

Predictability

As there is only one process there is little difficulty in ensuring regular execution and updates of output buffers.

Repeatability

Due to the sensitivity of the process execution to the timing of signals it is considered that it is more difficult to obtain "identical" conditions from run to run. If random input signals are applied to this type of system we are likely to get more variability in system behaviour and thus the will be a need for greater effort in interpretation of test results.

Functional Testability

In order to ascertain that the system has produced the correct transformation it is necessary to know the data upon which the function operated. It is more difficult to collect the required data for this type of system than it is for a type A system (see Traceability above)

Accessibility

As for a type A system.

4.3   Asynchronous Inputs with Multitasking System (Type C)

4.3.1 Description of behaviour

Figure 5 represents the operation of a multitasking system in which the input buffer may be updated during the execution of one or more processes. If the buffers and memory are well partitioned then this system can be made to approximate to that of the system represented in Figure 4. It may be possible to load a single process into the target during module testing in which case, the system, although it has the capabilities a represented in Figure 5, it is

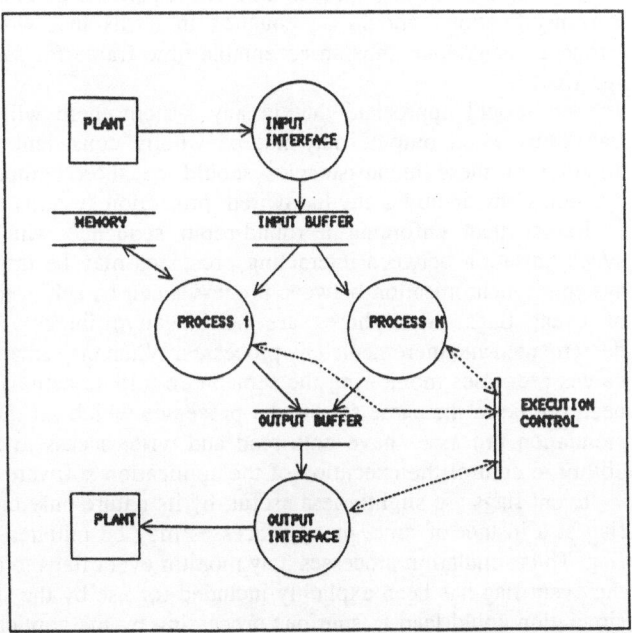

Figure 5 : Asynchronous Inputs in Multi-Tasking System

configured in the manner shown in Figure 4. If the software is coded so that internal copies of the input buffer are created within the process then with the provisions mentioned above, the system can be made to approximate to that in Figure 3.

The processes could be executed in a round-robin sequence, each process running to completion before handing back to the execution control. Each sequence could have

77

the same priority which would mean that the sequence of operation once defined would repeat ad infinitum. If each process could be assigned different priorities on line then the sequence of operation may alter; system operation would not be time deterministic. The principle of each process running to completion within a round robin sequence would help in testing because each process could be regarded as part of a greater overall process; the system, with all its tasks, would simplify to the system represented in Figure 4.

Consider a system in which the individual processes are allowed to free run without synchronisation. The processes may have different execution times and if there is no priority scheduling, the processes would cycle at different rates. This could create problems. Consider the situation in which process logic is fired by a transition in an input. "Fast" processes would execute initially "seeing" the transition and subsequently not "seeing" the transition, while slower processes would still be executing "seeing" the transition. This could lead to incorrect output patterns being applied until all processes had completed one cycle "seeing" the transition. If the test specification required that the tester observes the outputs continuously these patterns may be observed. Alternatively, if the system outputs are sampled some time after completion of the slow processes a wholly consistent set of outputs would be observed, failing to observe that transient (possibly dangerous) conditions had occurred due to multiple execution of the faster processes. Clearly such issues should be catered for in the system design. From the point of view of testing, the aim of which should be to discover faults [Mye76], the test specification should be couched in terms that would enable capture of any erroneous behaviour, thus an acceptable time frame for sampling test results must be specified.

One should appreciate that in any system there will be a short period during transitions when outputs may not be wholly consistent. As mentioned earlier, the duration of these inconsistencies should be short compared with associated plant dynamics, this includes any hardwired protection systems which monitor PES outputs.

Rather than enforcing a round-robin sequence which may be too restrictive, synchronisation between interacting processes may be implemented. In multi-tasking systems synchronisation between processes can be achieved using semaphores [All87] or event flags. Semaphores are non-negative integers which can be initialised, decremented and incremented by processes. When the semaphore reaches predetermined values processes monitoring the semaphore initiate actions. The monitoring processes need not be in the same set of the processes which set the semaphore value. Thus if simulation processes have both read and write access to semaphores then it has the ability to control the execution of the application software.

Event flags are slightly less useful; by its nature only one process may set an event flag at a instant of time, many processes may be initiated by the setting of the event flag. Thus simulation processes may monitor event flags to observe progress but, unless the event flag has been explicitly included for use by the simulation, its setting by the simulation could lead to spurious processing by the application software.

In Type C systems it is possible to allow the processes to run asynchronously. This is acceptable if the processes are well partitioned. Cost and compactness may be reasons for running independent processes in such an environment but from the safety point of view, if the processes are independent then running a system which introduces the possibility of interaction means that extra testing must be undertaken to demonstrate that no interaction takes places where none was intended. One way of reducing this testing overhead is to decompose the PES, allocating specific system functions to separate processors. There are a number of advantages in taking this approach:

. decomposition into safety-critical and non safety-critical functions;

. opportunity to use diverse hardware/system software/ application software;

. decomposition of the control problem into time critical and non-time critical operations.

The difficulties of testing a decomposed system include:

. the use of diverse hardware may introduce the need for diverse test environments;

. synchronisation of separate processor systems in establishing test pre-conditions.

Added difficulties of multi-tasking systems in general arise, even if synchronisation measures are implemented, due to asynchronous inputs affecting more than one process. If two or more processes use the same input then the instant of interrogation of the input buffer by each process and the time to cycle round each process become critical in assessing the execution of the code. If Process A uses input X at time t and Process B at time t+delta, then the opportunity for a data output mismatch arises if input X changes in the interval (t, t+delta). If the cycle time of Process A is Ta and Process Tb then the duration of the mismatch would be, at least the greater of Ta or Tb. This should be set against the alternative, which is that the inputs are latched until all processes complete their cycle. In this case both processes would produce signals which could be up to Ta or Tb out of date. The system should be designed such that this interval should be small compared with plant dynamics.

4.3.2 Discussion of Test Attributes

Controllability
As for type B systems with the addition of further processes to be controlled.

Observability
The time span over which the output buffer is changed may be increased due to the different execution times of the processes. Provided that the output interface effects all output changes in one scan then this type of system is no more difficult to deal with than with a type A or type B systems.

Traceability
As for type B systems.

Predictability
With the presence of multiple processes each with variable execution times, due to scheduling activities, it is fair to say that a type C is likely to be less predictable than type A or type B systems.

Repeatability

As for type B systems with the added complication of many processes to monitor and control.

Functional Testability

As for type B systems with the added complication of many processes, whose behaviour may have to be reproduced by the test environment.

Accessibility

With a multitasking system comes opportunity to use a further process to provide a simulator. The simulation code does not have to be embedded within or linked with code under test. If global memory areas have been employed in the application software, then declaration of the global data within the simulation code gives full access to the application data. These features make multi-tasking systems more accessible than single tasking systems (types A and B).

4.4   Requested I/O Operation (Type D)

4.4.1   Description of behaviour

Figure 6 represents a type of system in which the input and output signals are under the direct control of individual processes. In such a system the process may issue a request to the I/O sub-system of the target operating system, the execution of the process will then wait for a signal from the I/O sub-system. For an input request it returns a signal to the process indicating that the data requested is ready to be read. For an output request, the signal back from the I/O sub-system may merely be an acknowledge rather than a signal that the outputs change has been affected.

   This type of operation allows the process to obtain the most up to date information on plant status before execution of statements dependent on that information. The input/output data operations execute synchronously with the process code i.e. the process requests the information and waits for its return. On its return the process uses the data to complete its calculations. Using such a mechanism does not of course guarantee that the process executes synchronously with the plant. In a multitasking environment the time

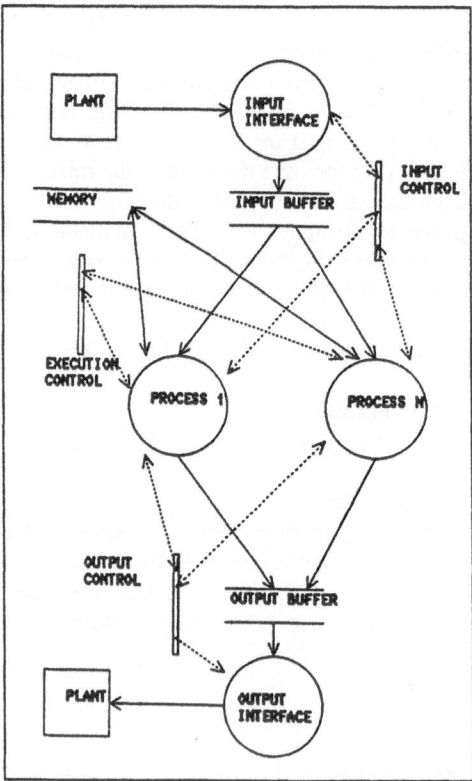

Figure 6 : Input/Output under direct control of Process

80

taken for each process to execute becomes indeterminate for a number of reasons which include: the various execution paths available within each process; the current status of the task, it may be waiting for an event; or the priority of the task compared with others waiting for the CPU.

4.4.2 Discussion of Testability Attributes

Controllability
As for type C systems.

Observability
As outputs may be effected by requests from individual processes, rather than from a single controlling process, the outputs may be changing at indeterminate points in time. Knowing when to look at outputs is important and thus there is the added complication of a signal being required by the monitor to indicate that an individual process has completed and that outputs related to the process may be recorded. Other outputs unrelated to the completed process may be in a state of flux due to the execution of other processes and thus should not be recorded. There is obviously a need for a more sophisticated monitor which can select and record only relevant signals.

Traceability
This type of system should be more traceable than type B or type C systems because the processes request input and wait for its arrival before continuing execution. Therefore we can observe the data areas at any time later sure in the knowledge that what we see now is the same data as the process logic used at some point earlier. Type B and type C systems allow data to be updated after processing possibly leading to confusion in interpreting results.

Repeatability
Because the requests for input may be distributed in time the systems behaviour becomes sensitive to timing of input signals than a type A system.

Functional Testability
As outputs are under direct control of each process they can be expected to change at any time the process requests output services. This means that constant monitoring of outputs would be required if evidence of attainment of response times are required.

Accessibility
As for type C systems.

5 Classification of PES Operation for Testability

Based on the discussions of earlier sections the attributes of each type of system have been assigned plus or minus marks relative to the other types. The sum of the plus and minus marks then form the score for the system. Such an analysis is to an extent subjective, but at least it is "observable".

It should be remembered that these scores relate only to testability attributes and not to the propriety of a system type for a particular safety critical application. There are many programming techniques and implementation features which may ameliorate some of the less desirable aspects of system operation, some of these techniques and features have been discussed however the scoring of the mode of operation of the has been done without the assumption that these defensive techniques have been implemented. On the basis of this analysis Type A systems are deemed the most "testable", this derives from their simplicity of operation and time deterministic behaviour.

Table 1 :   Classification of PES Operation by Testability Attributes

Attribute	Type A	Type B	Type C	Type D
Controllability	+	+	+	+
Observability	+	+	+	-
Traceability	+	-	-	+
Predictability	+	+	-	-
Repeatability	+	-	-	-
Functional Testability	+	-	-	-
Accessibility	-	-	+	+
Totals	+5	-1	-1	-1

6   Summary

One of the major problems in dynamic testing of systems is synchronisation:

. of input with code execution
. of output with code execution
. of code execution with monitors
. of processes with environment
. of processes with other processes

Whenever possible effort should be made to include in the design indicators which may be observed external to the processes under test. Independence of the implementation activities from test case generation may tend to exacerbate the lack of built-in instrumentation of systems. In time critical systems where additional software instrumentation cannot be tolerated the use of hardware monitors should be considered, see [CON92] for a discussion of relative merits.

In comparing four types of PES operation for testability, the simplest mode of operation emerged with the most favourable score.

However beyond the scoring system employed there are other considerations which were not felt to be quantifiable. Large general purpose operating systems are very complex items of software, timing and priority of tasks is difficult to establish and consequently synchronisation issues listed above are difficult to resolve. The complexity of the operating system tends to be unseen by the application designers and this can lead to misunderstanding about how the operating system will deal with requests of its services. General sentiment indicates that systems should be simple if they are to be testable, by selecting simple operating systems, and designing systems which recognise this simplicity, understanding will follow and unseen complexity will be avoided. In the context of high integrity systems the operating system should be configured so that timing can be guaranteed. This may involve segregation into separate PESs which provide dedicated functions and which have well established time deterministic behaviour.

It is worthwhile to note that drafts of future standards for system development indicate that software covered by the standards is likely to include operating systems, compilers and test routines as well as application programs. The requirements for testing such support software will be generally at the same level as the application software. The implication of this is that achieving the same coverage criteria for the operating system software as is required for the application software will be an enormous task for a large operating system, perhaps many orders of magnitude greater than that for the application software. Should such requirements survive to appear in authorized versions the pressure to employ simple operating systems may well become irresistible.

7   Acknowledgements

This work was partly funded by DTI as part of the CONTESSE project (IED4/1/9021).

8   References

[All87]   S.T. Allworth, R.N. Zobel: "Introduction to Real-time Software Design." Macmillan Education Ltd, 1987

[CON92]   CONTESSE Project Report : "Testability of the Design Features of Programmable Electronic Systems." DTI JFIT Ref :IED4/1/9021 NEI/Task_5/1 1992

[IEE88]   Institution of Electrical Engineers: "Guidelines for assuring testability." 1988

[Lew85]   Douglas Lewin: "Design of Logic Systems." Van Nostrand Reinhold (UK) Co. Ltd, 1985

[MOD91]   Ministry of Defence: " The Procurement of Safety Critical Software in Defence Equipment." DEF STAN 00-55/Issue 1 5 April 1991

[Mye76]   Glenford J. Myers: "Software Reliability." John Wiley & Sons, 1976

Data Management in Clinical Laboratory Information Systems

Richard Fink, Susan Oppert

Department of Clinical Biochemistry, West Middlesex University Hospital

Paul Collinson

Department of Clinical Biochemistry, May Day Hospital, Croydon

Glenn Cooke, Sukhdev Dhanjal

Safety Technology, Lloyd's Register, Croydon

Hamid Lesan, Roger Shaw

Applied Information Engineering, Lloyd's Register, Croydon

Abstract

This paper briefly reviews work undertaken within the DTI sponsored **MORSE** project. The Clinical Biochemistry Department of the West Middlesex University Hospital, one of the five project partners, provides clinical and laboratory services to a wide range of users. The Laboratory Information Management System used within the department has been developed using a range of commercially available hardware and software together with software that has been designed and developed within the laboratory. This paper reports on the first stages of safety analysis of the overall operations in the laboratory. This is a pre-cursor to the systematic re-development of the information system in the light of the findings of the safety analysis.

1   Introduction

Clinical Biochemistry is a branch of pathology in which constituents in body fluids are analysed in order to facilitate diagnosis, prognosis and monitoring of treatment. Virtually, all major hospitals contain a Clinical Biochemistry department, most of which provide a 24-hour service.

The recent history of the discipline is one of rapid development. Early analytical techniques were laborious and the workload was small, so results could be handwritten into ledgers, transcribed into patient reports and filed. Today, there is widespread use of automated analysers and an ever increasing number of tests are requested. Therefore computerised data management is essential to cope with the workload in an efficient and reliable manner. This is achieved at the Clinical Biochemistry department of West Middlesex Hospital using a system developed within the department.

Quality assurance of the analytical process is well established. However, in common with other safety related disciplines and applications, there is a concern about the reliability of computerised systems and the lack of generally accepted guidelines. These concerns are being addressed within the multi disciplinary DTI sponsored research project **MORSE**. West Middlesex hospital is one of the participants in this project.

2 The MORSE project

The last five years have seen the introduction of many new standards and guidelines directed at the developers of safety critical systems. Some of these standards aim to advise developers of safety critical systems containing programmable components on acceptable development practices in order to ensure that high standards of safety are maintained [MOD91b, MOD91a, IEC92a, IEC92b]. The software perspective that emerges from these studies is that there is a need to perform hazard analysis at the system level to determine the plausible failure modes and the potential hazards caused by these failures. The importance of the hazard analysis is that links are established between the unsafe consequences of failure and the plausible combinations of events leading to these consequences. Some of these failures represent failures of functions performed by the hardware and software of the system, others may be caused by operational errors.

One of the criticisms levelled at these standards is that they bring together a range of procedures, methods and design practices which, in combination, are essentially untried. It is claimed that more experience in their use is required before we can be sure that they address the problems of developing safe software. The **MORSE** project consortium (Dowty Controls, Lloyd's Register, Transmitton Ltd, West Middlesex University Hospital and the University of Cambridge) aims to address these concerns and in so doing produce a coherent approach aimed at the development of re-usable software for safety critical systems. The project builds on the achievements of the ESPRIT RAISE project [EP90, LG91, BG90] which has produced a tool supported formal language and method for software development. **MORSE** supports four themes of work:

Methodology: the application of specific methods, essentially formal in nature, to the development and safety analysis of safety critical software.

Design: the choice of appropriate engineering designs to support safety requirements.

Procedures: the procedural and management framework employed in the development process, with particular stress being given to the management and assessment of the safety case.

Case studies: the research performed on the above mentioned themes of the project will be influenced by and evaluated and used on three case studies. The case studies are:

- an aircraft landing gear control system,

Type of Analyser	Percentage		
	To LIMS	To Microcomputer	To both
Main Analyser	53	22	25
Supplementary Analyser	56	23	21
Emergency (Stat Analyser)	65	23	12
Endocrine/Immunoassay	38	44	18

Table 1: Interfacing of LIMS to laboratory equipment

- a system for monitoring and controlling a gas storage and distribution plant,
- a laboratory information management system.

This paper reports the work performed on the last case study which concerns an information system used in the West Middlesex Clinical Biochemistry laboratory. The case study aims to capture the requirements of the system and produce a formal specification. A safety analysis of the present system has been produced which forms the basis of this paper. In order to improve the integrity of the system, a number of areas will be chosen for re-implementation. This re-implementation will be performed in accordance with the methods and tools being developed by the project. Of incidental interest will be aspects of the human computer interface, and the procedures and working practices associated with the interface.

3 Laboratory information management systems

Computers have been used in laboratories since the mid-1960's for administrative and managerial support tasks. Recent developments have emphasised the direct interfacing of computers with laboratory instruments and the automatic performance of clerical tasks. In a recent survey of Laboratory Information Management Systems (LIMS) in use in the UK [ISC92], 78% of respondents stated that instruments had been interfaced with computers in their laboratories. A breakdown by analyser and computer type is shown in Table 1. Improvement in laboratory efficiency and saving of staff time were perceived as major benefits. Efficient data retrieval, improved quality control and a greater availability of information for administrative and management purposes are major facilities provided by LIMS.

The systems currently available commercially tend to be expensive, inflexible when in use, and difficult and costly to reconfigure or upgrade. This has led to a move towards a more flexible architecture based on off the shelf components. In the above mentioned survey the overall distribution of system types employed was as shown in Table 2. Although mainframe and mini-systems predominate there is a significant move towards PC based and PC networked systems. This is also reflected in the large number of MS-DOS users (35% of those surveyed). The range of hardware and software is large with over 35 different configurations reported.

System	Percentage
Mini	44%
Mainframe	21%
Micro network	18%
Stand–alone micro	17%

Table 2: Hardware configuration of LIMS

The development of LIMS have generally been undertaken in an unstructured manner without the use of formal system design methods or effective tools. Further, there are currently no standards set by UK, European or American regulatory authorities for development or assessment of LIMS beyond recommendations for software quality assurance such as ISO 9001 [ISO87].

4   The West Middlesex Hospital case study

The Clinical Biochemistry Department of the West Middlesex University Hospital (WMH) provides clinical and laboratory services for a wide spectrum of on–site and off–site users.

A summary block diagram of the laboratory computer network, as it pertains to the study, is shown in Figure 1. Patient samples are received in a reception area where they are given a unique laboratory reference number before they are prepared for analysis. Patient demographic and test details are entered into one of three terminals (DE) which are connected to a file server running the database and the network programs.

The file server (FS) is connected to a number of work–stations (WS), which are bilaterally interfaced to computers (AC) which in turn control large capacity (7000 tests/hr) analysers (A). Patient details including test request codes are transferred electronically from the work–stations to the analysers which perform the analyses required along with quality control checks before returning validated results back via the work–station and on to the file server hard disk (EHD). Further quality and validity checks are performed before test results are printed on hard copy for dispatch to clinical staff. The data is finally archived. A distinguishing feature of this system is the flexibility of retrieval and processing of stored data.

5   The safety analysis

As part of the WMH case study within the **MORSE** project a system level safety analysis was carried out on the data handling (electronic or otherwise) within the laboratory. The aim was to identify and analyse potential failures of the system which may adversely effect the integrity and timeliness of results leaving the laboratory. The recommendations resulting from the safety analysis cover aspects of laboratory design, operating procedures as well as improvements to the LIMS. The recommendations regarding the LIMS component of the system are to be implemented in subsequent stages of the **MORSE**

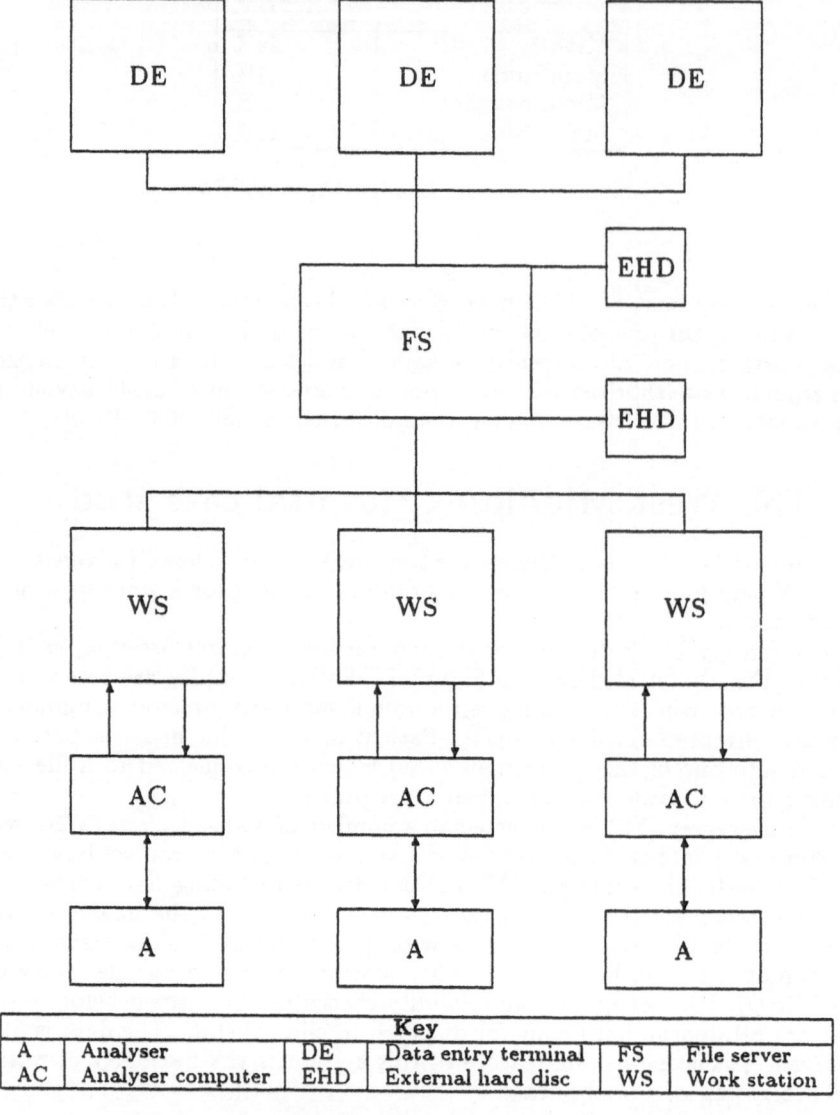

Key					
A	Analyser	DE	Data entry terminal	FS	File server
AC	Analyser computer	EHD	External hard disc	WS	Work station

Figure 1: The laboratory data management system

project.

This section discusses the hazard identification stage of the safety analysis which was carried out employing Failure Modes, Effects and Criticality Analysis (FMECA) and a Hazard and Operability (HAZOP) study. The initial aim of employing the two techniques was to compare the results obtained from each with a view to assessing their suitability for use in the laboratory.

A prerequisite for the safety analysis was a clear definition of the system, its boundaries and operations. This was produced in the form of a number of related modules (components or functional blocks) on a flow diagram which represented the main activities of the laboratory as follows:

1. flow of patient samples,

2. data flow diagrams,

3. hardware interconnections.

The diagrams were supported by text which described the sample preparation, the hardware used, the human computer interface, the database and associated programs and the computer/analyser interface. Each module was described in terms of its purpose, inputs and outputs.

It is important to note that the degree to which the system may be analysed is dependent on the amount of detail included in the system definition.

5.1 Failure Modes Effects and Criticality Analysis (FMECA)

FMECA [BSI91] is a qualitative reliability and safety analysis technique which focuses on the events that will cause unreliability or a hazard in a system and attempts to rank these events to reflect the seriousness of the consequent unreliability or hazard. The technique can be applied to a system which can be broken down into individual components. The components may be hardware or functional blocks as described within the system definition. During such an analysis the following questions should be answered for each component of the system:

1. How can the component fail?

2. What are the causes of the failure?

3. What are the consequences?

4. How critical are the consequences?

5. How often does the failure occur?

These aspects are recorded on a FMECA form specifically designed for the system in question.

The FMECA method requires the assessor to have a clear understanding of the function of each component along with all the associated inputs and outputs. The failure modes (deviations from design intent) are then investigated and the FMECA form is completed for each module. In this case the forms also

89

Integrity rating	(A)	Degree to which integrity of the data under consideration is lost.
Flow–rate rating	(B)	Delay to flow of data through the component.
System consequences rating	(C)	Likely effect on data leaving the overall system taking into account any recovery mechanism.
Failure rate	(D)	Frequency with which the failure is likely to occur.

Categories associated with ratings.

Integrity rating	0 : No loss of integrity. 1 : Partial loss of integrity. 2 : Total loss of integrity.
Flow–rate rating	0 : No delay. 1 : $\leq$ 3 hours delay. 2 : > 3 hours delay.
System consequences rating	0 : No consequences. 1 : Recovery possible. 2 : Recovery not possible.
Failure rate	1 : < 1 per month. 2 : Between 1 per month and 1 per week. 3 : > 1 per week.

Table 3: FMECA rating system

include columns for recording the criticality of the failure mode identified. The purpose of having a criticality rating scheme is to provide a systematic method of giving priorities to the large number of recommendations produced at the end of the analysis. The "appropriateness" of the rating can only be assessed through the use of engineering judgement. The criticality rating is a function of failure rate, system consequences rating, flow–rate rating and integrity rating. These are given numerical ranges as shown in Table 3.

In defining the criticality of an identified failure mode there was a debate as to whether this should be determined by considerations of patient management or more narrowly in terms of the performance of the laboratory. As the laboratory cannot always judge the importance to the doctor of an individual test result, it was decided to draw the boundary of the safety analysis around the department itself. This meant that the FMECA focused on safe handling of samples and data (both physical and electronic) within the laboratory system.

The effects on data of the failure mode of each component was scored in four ways as shown in Table 3. The total criticality rating (tcr) was calculated as follows,

$$tcr = (A + B + C) \times D$$

This formulae was then used to prioritise the hazards identified and to recommend improvements in the design and operation of the system.

5.2 Hazard and Operability (HAZOP) study

A hazard and operability study (HAZOP) [Kle86] is a structured approach to hazard identification. The study involves a systematic and methodical examination of design documents that describe a system. HAZOP is performed by a multidisciplinary team consisting of trained staff who are knowledgeable about the subject under study. The review is led by a chairperson who applies a number of guide words to each section of the system with the aim of establishing the ways in which that section can deviate from the intent of the system design.

The majority of HAZOP experience to date has been acquired in the process industries. In the WMH case study, guidewords had to be selected which were thought to be appropriate to the clinical laboratory environment. This matter is further discussed in Section 6.2.

6 Results of the safety analysis

The system definition of the WMH Biochemistry laboratory was produced in the form of sample flow diagrams, data flow diagrams, hardware interconnections and descriptive text. Each activity was reduced to small functional component, which in turn became the substrate for the FMECA and the HAZOP procedures.

In the following sections, examples of the application of FMECA and HAZOP to two functional components will be described. The first component involves sample handling, the second the performance of the computer system file server.

6.1 FMECA

Table 4 is an extract from the full FMECA that was undertaken and only shows details of the criticality analysis. In Example 1, sample tubes receive bar-code labels bearing critical information about the identity of the patient from whom the sample was drawn. Four failure modes are shown only the first of which will be explained in some detail as follows. If the wrong label is attached (failure mode) due to human error (failure cause), the sample will be wrongly identified (local consequences) and the incorrect result may be attributed to the patient (system consequences). The data is corrupted (integrity rating 2) but data flow through the component is unimpaired (flow–rate rating 0). The error may be identified at a later stage (system rating 1) and such problems occur approximately once a week (failure rate 3) resulting in a total rating of 9.

Example 2 refers to one of two external hard disks attached to the file server, for which three failure modes are shown (Table 4). The second entry is described as follows: if one hard disk fails, due to hardware error (failure cause), the affected drive cannot store data (local consequences) but since the disk is "mirrored" by an identical disc to which it is paired, information from the back up disc is used automatically. Consequently, there is no data corruption

Failure Mode	Integrity Rating	Flow–rate Rating	System Rating	Failure Rate	Total Rating
Example 1 – Attach bar code label to sample tube					
Wrong label attached.	2	0	1	3	9
Bar code printer failure.	0	1	1	1	2
No label attached. No number on tube.	2	0	1	2	6
No label attached. Number written on tube.	2	0	1	2	6
Example 2 – External hard disk					
Bad sectors.	0	0	0	1	0
Mechanical failure.	0	0	0	1	0
Driver board failure.	0	1	1	1	2

Table 4: FMECA – chosen criticality rating system

(integrity rating 0) and no impairment of data flow (flow rate rating 0). There are no system consequences and this failure occurs less than once per month (failure rate 1), giving a total rating of 0.

These examples appear to reflect engineering intuitions about the relative criticality of the failure modes discussed.

6.2 HAZOP

The guide words used for the HAZOP study on the computerised data management part of the laboratory were *no, not, more, less, as well as, part of, other than, sooner, later, where else, interrupt, reverse, more often, less often.* Table 5 shows the guide words indicating significant deviations from design intent with respect to the example described in Section 6.1.

Example 1 refers to the application of the bar code label to the patient sample tube. The word *no* referred to the absence of a label (deviation from design intent) due to human error (cause) with the result that the tube could not be processed on the analyser. *Other than* denoted the application of the incorrect label (deviation from design intent) with the result that the sample tube was attributed to the wrong patient. It was found that the application of guide words to activities was useful in stimulating the team to investigate deviations from design intent.

Example 2 focuses on one of the two external hard disk drives attached to the file server. *No*, in this context, produced three system consequences:

1. the system became entirely dependent on the working drive,

2. mirroring of the disks ceased,

3. the faulty drive required attention and maintenance.

Guide Words	Deviation	Cause	Consequences
Example 1 – (Reference 1.1.3)			
No	No label	Human error	Cannot process tube on analyser.
Other than	Wrong label	Human error	Sample id incorrect.
Later	No label	Handling error	Cannot process tube.
Example 2 – (Reference 4.2.3)			
No	No operating	Hardware failure	System reliant on one drive. No disk mirroring. Faulty drive needs attention.
As well	Both disks not operational	Hardware failure	File server shut down.

Table 5: Examples of HAZOP results

A recommendation was made that a simulation of this eventuality should be considered in order to test the actual behaviour and recovery procedures. *As well* introduced the possibility that both drives could fail simultaneously. This would result in a shut down of the entire network. A recommendation encouraging a review of preventive, recovery and backup procedures was made. These recommendations were written into the HAZOP records but have not been shown in Table 5 for lack of space.

7   Comment on the analysis - the laboratory perspective

For laboratory staff, the safety analysis was an interesting departure from the manner in which biochemical method comparisons are carried out; the performance of an analytical process is assessed against a reference method, the final arbiter being statistical rules which test whether the two techniques show agreement under various conditions. Such certitude is not available to the safety engineer, particularly when the subject matter involves the interaction of humans, computers and powerful analysers. Dealing with such problems requires a structured and methodical approach.

In retrospect, the yield of useful information from FMECA and HAZOP was thought to be sufficient to justify the considerable work involved. In particular, FMECA and HAZOP focused attention on every aspect of data management in the laboratory, whereas specific issues (eg hardware technology and similar concerns) tend to dominate the agenda when unstructured analysis of work practices are carried out. This and other advantages (see below) should be made apparent to laboratory staff embarking on safety analysis, as the process is very time consuming in the context of providing routine services in a busy and demanding hospital environment.

The FMECA provided a logical framework within which to work and produce useful results when considering machine and computer issues. For example, the possibility of a hard disk failure was placed in perspective by the zero rating which reflected the fault tolerance built into the system. Conversely when complex and interrelated procedures were considered, FMECA was less effective because of a tendency to compartmentalise the consideration of activities.

The HAZOP was popular with laboratory staff because it depends on an interactive approach which encourages participation and creativity. This freedom of discussion presents a potential problem in that the structure of the discussion may become compromised. Preventing this is in the remit of the adjudicator (chairperson of the session), who is a critical factor in the success or otherwise of the analysis. In general, the strength of the HAZOP study was its ability to cross activity boundaries and deal with interactions as well as issues pertaining to a single activity.

The output from HAZOP, even when expertly summarised and presented by a safety engineer, lacked the order of FMECA output in which a scoring system creates a prioritised list of problems which could be of use to a laboratory manager in a decision making process. This problem could be addressed for HAZOP's provided that the criteria for setting the priority can be clearly stated. In further contrasting the two methods, there is some truth in the assertion that FMECA is useful when investigating single components of a system since this can be done by individuals. However HAZOP is more useful in investigating interactions between system components although it requires a team of personnel to take part in the study. This division is probably simplistic and the ideal technique as far as a laboratory is concerned probably requires a contribution from both methods.

8 Discussion and conclusions

The advent of high capacity analysers which produce large quantities of patient test results has stimulated an interest in the safety aspects of laboratory data management systems. The application of techniques such as HAZOP and FMECA has proved very instructive in that a better appreciation has been gained of the hazards that can befall data as it makes the journey from human to computer, computer to analyser and analyser by a circuitous route, back to the human again. As a result of applying these procedures problems have been identified and appropriate changes made. The LIMS related problems will be addressed in the next stage of the project.

These problems can be minimised if they are identified in the planning stage of a laboratory service, regularly monitored once the laboratory is in full scale production, and particularly during the management of change. To achieve these objectives, a method is required that is user friendly but not too simplistic, able to assimilate complex and interrelated activities while retaining a sensitivity to important detail, and finally one that provides information in a systematic way that is prioritised by an intelligible scoring system.

9 The next step

The next stage in this case study will involve the prioritisation and selection of safety related components of the system in particular those supported by software. These will be formally specified and developed using the RAISE method and basic safety management procedures.

References

[BG90]   S. Brock. and C. W. George. RAISE method manual. Technical Report LACOS/CRI/DOC/3/v1, CRI, 1990.

[BSI91]   BSI. *BS 5760 : Reliability of Systems, equipment and components Part 5. Guide to failure modes, effects and criticality analysis (FMEA and FMECA)*. British Standards Institute, 1991.

[EP90]   K. E. Erikson and S. Prehn. RAISE overview. Technical Report LACOS/CRI/DOC/9/v2, CRI, June 1990.

[IEC92a] IEC/TC65A(Secretariat)122. *Software for Computers in the Application of Industrial Safety-Related Systems*. Geneva, 1992.

[IEC92b] IEC/TC65A(Secretariat)123.
          Functional Safety of Electrical/Electronic/Programmable Electronic Systems: Generic Aspects. Part 1 : General Requirements. Geneva, 1992.

[ISC92]   S. Ike, P. Smith, and P. O. Collinson. Laboratory information management systems – the state of the art. To be published in the Annals of Clinical BioChemistry, 1992.

[ISO87]   ISO 9001: Quality systems – model for quality assurance in design/development, production, installation and servicing. International Standards Organisation, 1987.

[Kle86]   T. A. Kletz. *HAZOP and HAZAN Notes on the Identification and Assessment of Hazards*. The Institute of Chemical Engineers, 1986.

[LG91]   The RAISE Language Group. *The RAISE Specification Language*. Prentice Hall, 1991.

[MOD91a] Ministry Of Defence. *Hazard Analysis and Safety Classification of the Computer and Programmable Electronic System Elements of Defence Equipment – Interim Defence Standard 00-56/1*. Ministry of Defence, April 1991.

[MOD91b] Ministry Of Defence. *The Procurement of Safety Critical Software in Defence Equipment – Interim Defence Standard 00-55/1*. Ministry of Defence, April 1991.

Software Engineering Methods for Industrial Safety Related Applications

Audrey Canning
ERA Technology Ltd
Leatherhead, UK

Extract

This paper is based on the work undertaken during the first year of a four year collaborative project entitled "Software Engineering Methods for Safe Programmable Logic Controllers". The project is part of the joint DTI/Serc "Safety Critical Systems Initiative". In the paper the background to the project is presented, and the principal objectives of the project are summarised. Whilst it is too early to present detailed results, the major issues are summarised. The principal conclusion to date is that the cross-sectoral background of project members has challenged many preconceived notions.

1 Introduction

Many practical industrial systems are based currently on Programmable Controller or Programmable Logic Controller (PLC) systems. These systems have significant advantages for safety. For example, off-the-shelf, industrially hardened components are used which can be considered to be proven through a wide industrial market base. Specialist re-usable software functions are likely to be available, including self diagnostics, networking and process control functions. Application specific functions are achieved through use of a language which is familiar to the process design engineer.

One of the most widely used application languages for PLC systems is known as "ladder logic". Intended to represent the coil and contacts of electronic relay circuits, in fact the language contains similar logical functions to those available in microprocessor assembler languages. It is now recognized that the use of ladder logic, although heavily favoured by process engineers as being easily understandable and familiar, produces software which:

(a)   suffers from a lack of high level requirements analysis

(b)   does not truly represent the parallel nature of the underlying "relay" model

(c)   is difficult to review and debug since the code is dispersed as small elements through the program

(d)   lacks facilities for evaluating the effects of timers, counters, scan times and other logic constructs which may cause the system to enter invalid states.

As a result PLC based systems frequently suffer from software which behaves in unexpected ways, and as a result of time and cost overruns in development.

The use of suitable software engineering methods plays a significant role in producing software which is reasonably free from error, fulfils the customer requirements, can be easily maintained, and meets development time and cost constraints. There is wide variation, however, in the software development methods in use throughout UK industry. Moreover, there are at present few guide-lines for determining how well a set of methods will perform in a particular application.

Even when suitable methods are available the complexity of the software design activity may be so great that the unaided application of the methods may not be successful. In fact some standards (e.g. Def Stan 00-55) now mandate the use of automated support tools, such as compilers, configuration management, and verification tools, in order to reduce the number of errors made by design staff. A drawback of such tools is that their complexity can itself result in problems in application of the tool.

2   The SEMSPLC Project

In view of the link between "good" methods and high quality software, a collaborative project[1] has been set up to investigate the extent to which existing software engineering methods can be applied in industrial applications. The project has the objectives of:

---

1 Software Engineering Methods for Safe Programmable Logic Controllers (SEMSPLC)
Part of the joint DTI/Serc Safety Critical Systems Initiative
Project partners are: ERA Technology Ltd, ICL Ltd, Liverpool Data Research Associates Ltd, British Gas plc, Servelec Ltd and University of York

(a)   establishing a software engineering method suitable for the development of safety critical PLC based programs in an application language suited to industrial control

(b)   defining conformance standards which could support the application of the method.

In order to provide a qualitative measure of the success of the above objectives, the project has the subsidiary objective of attempting to identify comparative measures of safety levels achieved and the costs involved in the use of software engineering methods and tools.

Currently, the project is nearing completion of the first year of a four year work programme. During the year, studies have been carried out into the nature of the PLC design process, the theoretical processes involved in system and software engineering, and the needs of industrial method and tool users. Other activities have been carried out to investigate the commercial and safety impact of re-using software, and to review candidate metrics. The results of the studies are being analysed to derive a more detailed set of requirements for the later phases, to identify the activities which need to be addressed by the proposed method, and to identify procedures and notations which may comprise the elements of the method.

During Phase 2, due to commence in December 1992, the information collated during Phase I will be used for three purposes:

(a)   to define requirements analysis, verification and validation methods for use during PLC software development which can be used in conjunction with existing PLC languages

(b)   to specify conformance criteria for tools which could automate the requirements capture, validation and verification stages of the PLC software development process

(c)   to establish a means of collecting software safety metrics.

During the third phase, the PLC methods are to be tested on demonstrator projects with the objective of evaluating and refining the methods. In addition, the conformance criteria are to be verified through prototype development of selected tools (ie requirements capture and verification) and comparison of the resulting tools with conventional tools.

At the time of writing a total of fourteen reports have been prepared by the project partners, and, in addition, a number of interim working papers have been submitted. The systematic analysis of the results of these studies in still in progress, and hence it is not possible to provide the detailed findings of the work. It is clear from the reviews that have taken place, that each partner has been exposed to new ideas and approaches, although the benefit of the application of the ideas to industrial use is not necessarily apparent. Some of the important issues which have arisen during the course of the work are discussed below.

The Characteristic Features of Industrial PLCs

The scope of the project was defined in the original work plan as limited to the development of methods for PLCs. The definition of a PLC, for the partners coming from the software engineering background, has been difficult to grasp. In particular, the PLC is not characterised by the software language used, since international standards in this area recognise that a PLC may support a wide variety of different language types, including special application languages and high level languages. Neither is the definition characterised by the basic hardware processor since PLC architectures can be widely dissimilar.

In contrast, the industrial partners have a common understanding of a PLC, based on the characteristics of the application environment. In addition to the hardware characteristics, which will be appropriate for use in the industrial environment, the PLC is agreed to be a computer based system, which has been optimised to carry out industrial control functions, through provision of specialised application support functions and facilities, and through the ability to cater for large numbers of input and output signals. Differences exist in terms of the type of industrial control algorithm implemented by the PLC (in particular, whether the algorithm is based on batch control or sequential control) and in response times (for example whether the application is concerned with machine control or chemical processes). The SEMSPLC project includes sponsors with applications in each of these areas and, as a result, is likely to require a number of different types of development method.

The concept of the method to be provided by the project has evolved as the project has proceeded. In the work plan, the definition of a software engineering method was viewed as covering the full technical development lifecycle, from requirements capture to system validation. As the project has progressed, it has become apparent that the method will need to address not just technical issues, but aspects such as maintenance and management. Typical of such issues is the problem of subcontracting parts of the development process. The SEMSPLC partners have experience that, as a project proceeds and more detailed design is considered, subcontractors may uncover safety issues that were not considered during the initial systems concept (and costing). It may be necessary to consider special inter-locking arrangements for example, which may in turn have an impact on other parts of the system design, such as the start up procedures. The responsibility for undertaking the necessary additional work and for evaluating the effects on the overall design may not be clearly defined, and contractual relationships may well obscure the technical needs.

Safety as a Property of the System or of the Software

Some discussion has been generated with regard to the term "Safety". "Safety" for the industrial partners involves the concept of injury, mainly to humans, but also, potentially to the environment. Failure modes and effects analyses on a variety of system models have been initiated to identify the way in which a system can lead to injury. The basis for this work is that failure modes and effects analysis need not, as is currently prevalent, be applied only to physical and component models of a system, but instead can be used in conjunction with software models to determine a qualitative assessment on the effects on human safety.

In the context of software, however, a system is dangerous only in the sense that information can be released, or fail to be released, from the system and can stimulate a hazard in the external environment. For some partners, the concept of assuring the integrity of information entering and leaving the system is a well established technology, and is based on mathematical relationships between the output information, the input information and the internal states of the software system. There is an alternative view therefore that software safety is concerned more with the information integrity, than with the wider human and environmental impacts.

Interpretation of Graphical Models

One of the interesting results which has been identified as a result of the diverse backgrounds of the SEMSPLC partners, is the ease with which graphical models can be interpreted in different ways, even within the same community. One advantage of the ladder logic language was believed to be that the language models the effects of hardware relay logic and, by implication, represents a parallel model of the system behaviour. It was believed that the act of interpreting the parallel model in a computer system led to difficulties, in as much as true parallelism could not be supported. Discussions with the users have confirmed that this is the process engineer's interpretation; however, discussions with the systems builders show that long familiarity with the sequential operation of a computer system leads the development engineers to interpret the ladder logic in sequential fashion. The moral for the design of safety critical systems is not to make any assumptions as to another's understanding of the written specification, since that specification could well be interpreted in the light of the experience and training of a particular individual!

Timing Issues

From the outset, it was recognised that ladder logic programs work well for static logic algorithms, but tend not to support complex timing relationships. For example, it is clear that the PLC, by providing a scan based operating kernel, provides the application engineers with an ideal environment for the design of synchronous logic systems. Providing the necessary control algorithms can be constrained to complete within the PLC scan time, and the individual ladder rungs are treated as series logic elements, then the PLC programming approach does constrain the program to single thread execution and, as such, is commended as a basis for safe design. Difficulties can arise, however, since the development engineer is not constrained to use the PLC in this way, and can and does introduce dependencies between scans which corrupt the sequential behaviour of the logic.

Re-use as a Safety Argument

One other area which is causing some discussion, is the findings from the re-use study. For this study two activities were carried out, a literature review was undertaken to establish the major issues surrounding re-use and a survey a SEMSPLC project members and industry at large was also undertaken. The survey results raise two issues.

Firstly, the results are at variance with the literature as regards the "Not Invented Here" (NIH) syndrome. In particular, the majority of the industrial respondents appear to welcome the idea of software re-use, and many declare that they are already making much use of re-usable code and designs. Since there is evidence that significant cost savings can be made through re-use, perhaps NIH is less significant where performance is monitored by budget as well as by technical innovation!

The second issue arising from the study is that little evidence could be found that re-use decreases the error rate of software, or increases the safety of the software. Whilst this conclusion is based more on qualitative information than on scientific evidence, it could have major implications for the construction of many industrial safety cases if the results were confirmed on a wider basis. In particular, a safety case built on the use of "well-proven" software would be difficult to justify!

4   Concluding Remarks

In undertaking the SEMSPLC work programme, it was recognised from the outset that the enabling technologies essentially exist in the fields of industrial process control, systems design and software engineering. For example "ad-hoc" methods are in use throughout industry, which clearly do provide systems that fulfil their objectives.

The innovative content of the work programme was recognised to arise from the intent to bring together widely differing approaches to computer systems engineering so as to gain valuable insights into the software engineering discipline as a whole. The issues discussed above have arisen directly as a result of the need to understand the differing opinions of the project members. From the nature of these issues, it can be seen that the wide range of backgrounds has lead to many preconceived notions being challenged. It is concluded that the early work provides a firm basis on which to move forward to gain an increased understanding of the problems of developing software for industrial applications.

Finite Element Safety Critical Software

by A.J.Morris

College of Aeronautics

Cranfield Institute of Technology

Cranfield, Bedford

on behalf of the SAFESA Project

CIT

Nuclear Electric

Lloyd's Register

W.S.Atkins

Assessment Services

1. General Overview of Analysis Evaluation

1.1 Introduction:

The purpose of this paper is to describe the first stages in generating a methodology for controlling the effect of errors when a structure is analysed by the Finite Element method associated with the work of the SAFESA project. The background to this work is the desire to qualify safety critical structures by finite element analysis only. In certain cases the emphasis is to remove or decrease the role of testing in the certification procedures, for example in aircraft construction. In other cases it is to confront the problem of predicting the behaviour of a structural system where test is impossible, in such cases as the design of stable long term nuclear storage systems. The central focci of the work of SAFESA are that the structures being designed are safety critical and that they are to be analysed by the Finite Element Method. In this situation the use of the Finite Element Method can be considered as an application of a safety critical software. We are, therefore, attempting to control errors in the analysis process so that a prescribed confidence limit can be ascribed to the eventual analysis results.

In this context the process is much wider than simply controlling errors associated with the basic FE software. Rather it is concerned with errors originating from the total use of the Finite Element Method in the analysis of structures at the design stage. These cover a vast range stretching through modelling considerations, user errors, distorted meshes, inappropriate elements etc which extend well beyond the errors encountered in conventional software error analysis. Nevertheless, the methods used in the development of software systems to control errors are important in providing an outline methodology which can be exploited to control errors in the total analysis process.

In order to quantify and assess the effects of errors on the results from an analysis the process itself must be understood and also the errors identified. The linking of errors to the actual finite element analysis can then be attempted. Thus, in the sequel, we start by reviewing some of the approaches used in software development which have a bearing on the F.E. analysis process. The paper then moves on to considering more specific F.E. aspects relating to hierarchical concepts and error taxonomy.

Specifically sections 2 and 3 examine the decomposition of errors and the analysis whilst section 4 then classifies the errors into a taxonomy which is linked to the analysis. It should be emphasised that the paper represents an overview of the work of the SAFESA project which is still being researched. Nevertheless, more details of the ideas presented herein are available and can be found in the project documents.

Until recently, the standard software development methodologies consisted of a requirement phase, a functional and detail design phase and a verification phase, but without a separately identifiable effort devoted to safety considerations. Several factors, in recent years, have changed this view forcing a focused and specific effort on safety aspects throughout the development process. One important aspect is the use of safety critical systems which are controlled by software and where it is vital to be able to ascertain that this type of software does not contain an error that will result in human injury or loss of life.

In the creation of applications software much of the endeavour was (and still is) directed toward developing software to the required quality, cost and timescales. However, with safety critical software, an additional requirement is to demonstrate that this has been done and quality standards achieved. A significant part of this demonstration is by verification of the software following development. Formal testing techniques exist today to prove that the software did meet its specifications. However, if the software is not performing according to the expected function there is a need to detect the error which caused the deviation. A great deal of effort has been deployed in order to devise methodologies that enable the minimisation of the software faults. The classification of the types of errors into some sort of taxonomy is widely accepted as the starting point for the determination of the error's predisposing factors.

1.2 Decomposition

The first stage in any error assessment or control procedure is the decomposition of the processes into basic building blocks. In the present paper this requires decompositing both the errors themselves and the actual analysis process. Section 2 addresses the problem from the viewpoint that errors need to be linked to the analysis in a way which allows their effects to be detected and their influence traced. This leads to the concept of generating a tracing process which is associated with specific error types. Such a tracing process is not a true decomposition but is a useful tool in locating errors in a specific part of the analysis task.

In order to map the tracing process onto the analysis procedure two tasks need to be performed in section 3. The first is to decompose the analysis itself into a series of processes or phases in order that the first stage of the tracing process can be performed. The second is to provide a taxonomy of the errors associated with each phase of the analysis process.

The full error analysis process requires that the decomposed analysis process has each phase fully defined in terms of the errors associated with it and that each error, itself, is fully specified. At the present stage of the SAFESA project this complex process of definition and linkage is far from complete. Nevertheless, some progress has been made and in the later parts of this paper the work completed to date by the project is reported.

1.3 The Use of Taxonomy Codes in Error Classification

Once a basic decomposition is available a return can be made to decomposing errors which, in effect, implies that some form of categorisation is available. In other words, a comprehensive error categorisation is a necessary prerequisite to determine the possible errors which can arise in an analysis. The development of error taxonomies has occupied a number of researchers (e.g. Beizer [2], Schneidewind & Hoffmann [5], Endres [6], Morris [1], Nakajo & Kume [7], etc...) who have recommended, or actually used, error categorisation as a base for the testing technique they advocate.

Unfortunately, there does not seem to exist a universally correct way to categorize errors. The latter are often quantified from different viewpoints depending on the goals of the error analysis. Hence, a given error can be put in one category or another depending on the logic followed (or the criteria used) in the classification process. For example, what is viewed by an operator as a pure software failure can be referred to as a user carelessness by the software programmer. Therefore, it may be useful to note that in this situation the endeavour should be directed to devise a useful taxonomy, in order to provide a basis for error assessment, rather than to develop the 'right' taxonomy which may never be achieved.

Beizer [2], for instance, developed a detailed classification of errors with a taxonomy code for each error. However, and despite the level of details with which the errors are categorised, Beizer's hierarchical classification is not claimed to be a universal one which can be used for any type of software evaluation (statistical experiments have shown that at least 4.7% of the reported errors were not covered by the classification). The only argument raised in favour of this particular classification (rather than adopting the IEEE draft standard on error codification) is simply that it covers the territory.

1.4 Error Control

The sorting of errors into general taxonomic categories is one valuable step in the understanding of errors, but it still does not take us very far. The software engineering discipline has expended a great deal of effort in order to take the necessary measures to minimise the impact of these errors. In general, one of the following three options is used:

Error Avoidance:

Most of the effort is devoted to avoid introducing errors (i.e. prevent their initial insertion). Techniques such as modularisation are used to avoid errors.

Error Elimination:

This is the process by which errors are removed from the developed system. This process is accomplished through reviews, analyses and the application of various testing methods.

Error Tolerance:

Error tolerance is the application of design, testing and coding techniques that will minimise the potential impact of errors on the operation of the system. In other words, an error is precluded from manifesting itself in an undesirable or unsafe manner.

Note that for the SAFESA project a hybrid approach which combines the three options above is considered appropriate, especially that the project deals with structural design software which is safety critical and where any error is a potential life threat. But as outlined earlier, the efficiency of any of the above options depends considerably on how much one knows about the errors that need to be avoided, eliminated or tolerated.

2 Error Decomposition

2.1 Consistency

The hierarchical decomposition of the errors associated with the analysis of a structure by the Finite Element Method requires an examination of the analysis process itself and the error classes. The association with the analysis process implies that the errors relate to the two basic stages of a finite element analysis. These two stages are concerned with the overall consistency of the method which break down into internal and external consistency.

Internal consistency relates to the errors which occur as the F.E Method attempts to solve a continuum mechanics problem. This is an ideal problem for which there is a valid solution. Error control in this phase of the analysis process is concerned with assuring that the computer code produces the correct solution to the ideal continuum problem and is properly called validation i.e. assuring that the code gives rise to a solution which is mathematically valid.

External consistency relates to the problem of assuring that the ideal continuum problem properly represents the real world structure which is being modelled. This phase is concerned with assuring that a solution from an analysis actually corresponds to the behaviour of the real world structure. In the design process the difficulties associated with this aspect of the error control process are compounded by the fact that the real world structure often exists only in the form of a set of drawings or CAD data forms. The trapping and control of errors at this stage are really associated with verifying the analysis and, thus, the process is know as verification.

2.2 Error Traceability

Ideally the decomposition process should lead to the creation of a tracing process whereby errors associated with internal and external consistency can be detected and their influence assessed. An outline of such a process is shown in tree 1 where the analysis process is assumed to be decomposed into a set of processes or phases. Associated with each phase are a set of errors which require categorising into a set of types. This process of categorisation is discussed in section 4 and is reported in greater detail in a SAFESA report 9034/TR/CIT/2001/0.0/15.9.92. The context is wider than simply controlling errors associated with the basic F.E. software. Rather it is concerned with errors originating from the total use of the Finite Element Method in the analysis of structures and in this context is associated with both internal and external consistency which relate to specific parts of the analysis process in a quasi decomposed manner.

The tracing procedure is also concerned with the ability of the analysis process to both detect an error and to follow any propagation path. Assuming that we have

108

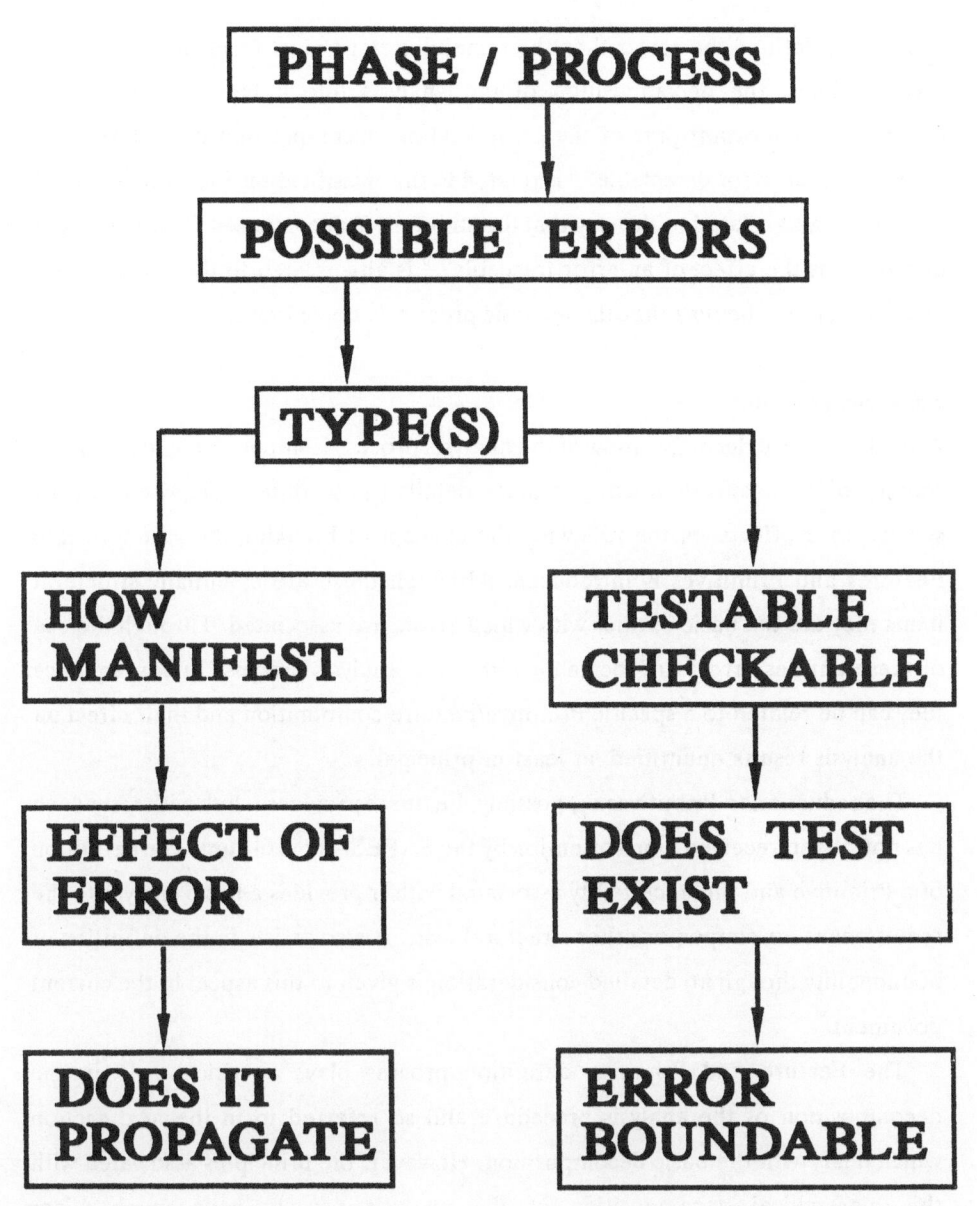

Tree 1   Error   Trace

adequately defined the errors then the remaining stage must focus on the analysis process. Thus, the decomposition of the analysis into a hierarchical process becomes an important part of the error analysis decomposition procedure. The question " is an error detectable? " is related to the classification question discussed later but is also related to the way that the structure is decomposed. The associated question " is the effect of an error traceable? " is also related to the classification problem and to the way that the analysis process is decomposed.

2.3 Error Location

Although a basic decomposition of the analysis process is important in locating the sources of error and their effect a more detailed procedure is required to fully specify error effects. In the following the concept of breaking the structure into Features and Primitives is introduced. Although these are essentially structural items they are the basic entities with which errors are associated. Through the use of these entities errors are locatable within the analysis process. In consequence they can be related to a specific Primitive/Feature combination and their effect on the analysis results quantified, at least in principal.

The reduction to Primitives or, possibly, Features provides a link with test which has not, as yet, received much attention by the SAFESA consortium. The definition of a Primitive and the uncertainty associated with it provides a basic entry into the requirement for coupon or other structural tests. It also assists in the definition of boundability though no detailed consideration is given to this aspect in the current document.

The Feature and Primitive definition process plays a critical role in the decomposition of the analysis procedure and so referred to in the next section which deals with the basic decomposition. However, the principles associated with this hierarchical decomposition of the analysis process have much wider implications which are not dealt with here. It should be noted that this particular decomposition procedure is, in fact, simply part of the current move to decompose processes into Object Oriented data representations.

3 Analysis Decomposition

3.1 Real World Decomposition

The initial process in the decomposition of a structural analysis problem is the definition of the actual Real World Problem itself. This does not imply any attempt to generate the idealised model which is employed by the analysis software to find a numerical solution. Rather it is the answer to the question of what is the actual problem being solved. The requirement is for an initial decomposition which fully defines the Real World Problem to be analysed.

In essence this decomposition is equivalent to defining a taxonomy for the variety of factors which make up the Real World Problem. Once this has been performed a second, and more complex, decomposition can be entered which relates to the analysis process associated with solving this problem and which constitutes the remaining part of the threads analysis discussed in section 3.3.

Defining the Real World Problem requires an assessment of the parameters which provide the basic description of the problem which is to be analysed by the Finite Element Method. This assessment is strongly linked to the requirements of any qualification procedure imposed on the design. Such factors as 'the purpose of the design', 'the limiting design criteria' etc. are clearly a function of the process which will lead to the eventual structure being qualified or certified. In order to accommodate these aspects some form of initial decomposition of the Real World Problem is required which breaks down the problem into terms accessible to the analysis process as a basic starting point.

Because this initial decomposition is so strongly linked to a specific industrial application and its associated qualification procedure it is not possible to provide an industry independent process. Nevertheless, a description is required of the major controlling parameters which make up the definition of the Real World Problem, such as loads, design critical factors etc. When these factors are examined it is clear that the process is one of finding a taxonomy which adequately defines the terms being addressed. To illustrate the points being made the specific case of deriving a basic definition of the loads is now considered.

The process of defining the loading actions required in Step 1 is clearly a complex process and depends upon the industrial base being addressed. Whilst a general outline taxonomy has been developed by the SAFESA project the approach adopted in this paper is to focus on the construction industry as typical of a major application area.

Before specifying a loads taxonomy for insertion into the threads analysis it is apparent that "loads" needs to be defined and perhaps broken down as follows:

i.   Identification of failure modes leads to identification of DESIGN SITUATIONS and corresponding DESIGN LOAD CASES;

ii.   For each design situation all the CONTRIBUTING LOADS need to be identified in each DESIGN LOAD CASE;

iii.   The MAGNITUDE OF LOAD then needs to be identified;

iv.   The appropriate PARTIAL LOAD FACTOR then needs to be identified (dependent on the type of analysis).

v.   The appropriate PARTIAL RESISTANCE (MATERIAL) FACTOR then needs to be identified.

vi.   Finally FACTORED LOADS need to be compared against FACTORED RESISTANCES to ensure structural qualification.

With these defining factors in mind two taxonomies relating to the definition of loading actions are given in loads diagrams 1 and 2.

A similar process can be applied to all the other aspects associated with the design environment until an adequate taxonometric description of the Real World Problem is achieved. At this stage we are ready to enter the threads analysis of the next section which addresses the real decomposition of the finite element process.

3.2 Threads

Section 2 described the linkage of the error control process with the need to decompose the analysis process. In approaching this problem consideration was given to the classical software decomposition process where a given computer

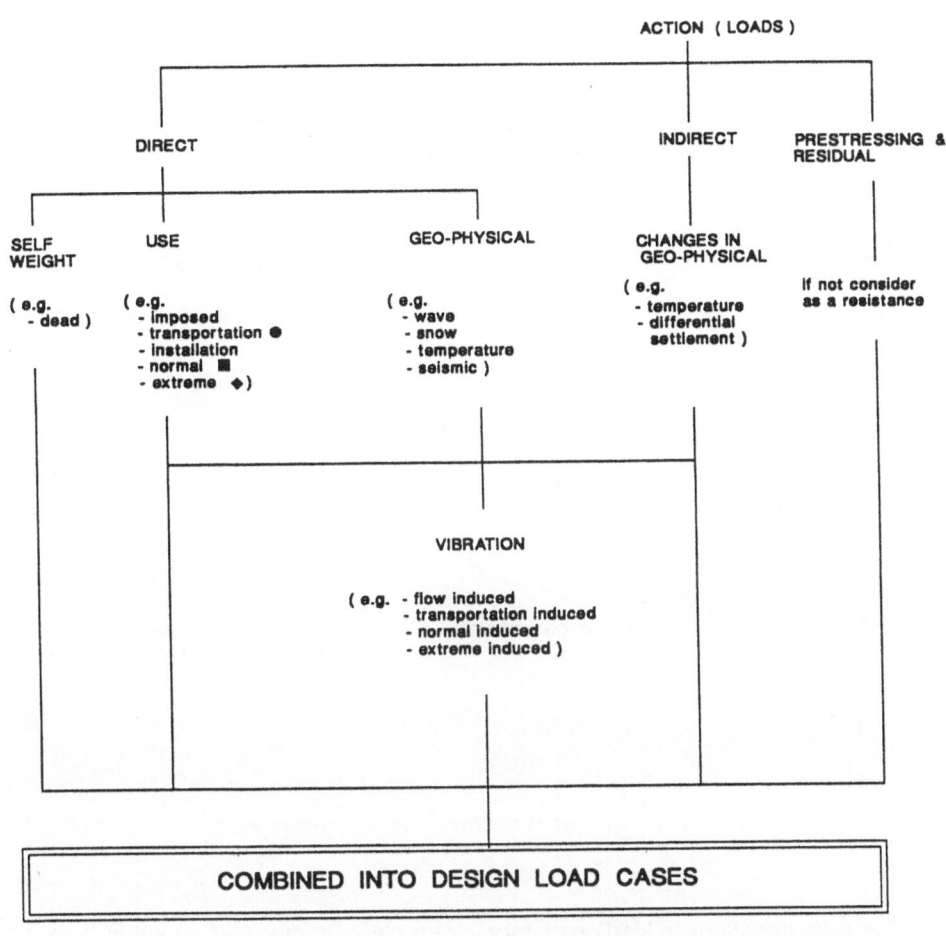

ACTION (LOADS)

DIRECT

INDIRECT

PRESTRESSING &
RESIDUAL

SELF
WEIGHT

USE

GEO-PHYSICAL

CHANGES IN
GEO-PHYSICAL

(e.g.
- dead)

(e.g.
- imposed
- transportation ●
- installation
- normal ■
- extreme ◆)

(e.g.
- wave
- snow
- temperature
- seismic)

(e.g.
- temperature
- differential
settlement)

If not consider
as a resistance

VIBRATION

(e.g. - flow induced
- transportation induced
- normal induced
- extreme induced)

COMBINED INTO DESIGN LOAD CASES

Notes   ● Transportation   by road, rail, sea, air

  ■ Normal   body forces / pressures (can also apply to GEO-PHYSICAL)

  ◆ Extreme   blast, shock, impact (can also apply to GEO-PHYSICAL)

Also   Could classify: DIRECT (SELF WEIGHT) &
PRESTRESSING / RESIDUAL as STATIC
INDIRECT & DIRECT (OTHERS) as DYNAMIC

Also   Could address: TEMPORAL NATURE
SPATIAL NATURE

DIAGRAM 1   CONSTRUCTION INDUSTRY EXAMPLE

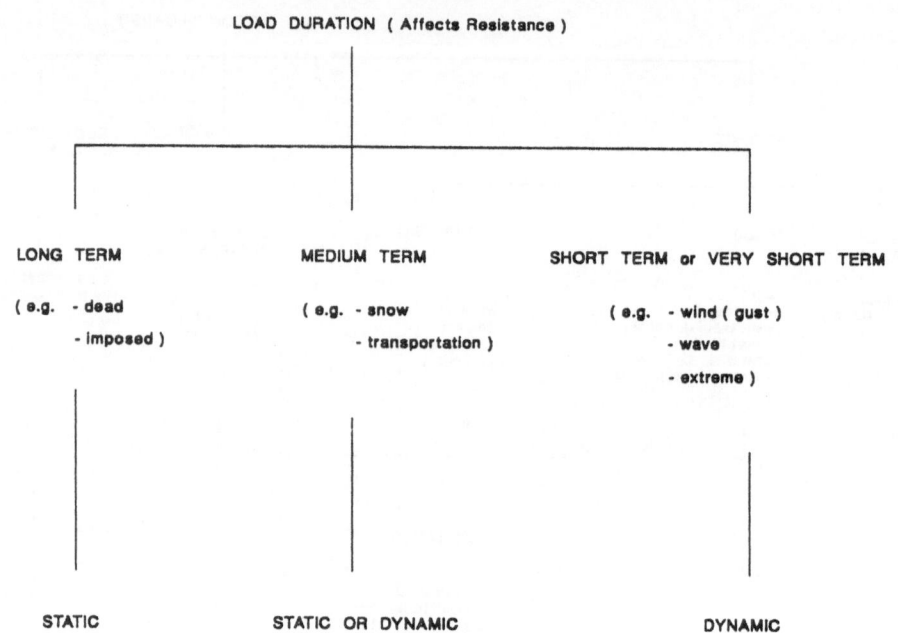

LOAD DURATION (Affects Resistance)

LONG TERM	MEDIUM TERM	SHORT TERM or VERY SHORT TERM
(e.g. - dead	(e.g. - snow	(e.g. - wind (gust)
- imposed)	- transportation)	- wave
		- extreme)

| STATIC | STATIC OR DYNAMIC | DYNAMIC |

Notes

i. Temporal variations in loading history - for a given loading history either a "static" or a "dynamic" analysis may be appropriate depending on the individual nature of the structure (e.g. a 15 second wind gust on a cladding panel might be satisfactorily treated as static whereas a lightweight mast might need to be analysed dynamically).

ii. Although a structure responds dynamically it can, in some circumstances, be admissible to qualify structures by "quasi-static" criteria (e.g. seismic loading).

DIAGRAM 2   TEMPORAL ASPECTS

package is broken down in the form of a process chart. In this approach the input/output procedures for each major part of the program are indentified and their influence assessed.

The associated process for the Finite Element Method is slightly different and in consequence is referred to in the present document as a threads analysis. The process requires breaking down the analysis into a set of stages and for each stage applying an algorithm which identifies discrete steps together with inputs/outputs. Table 1 shows how the threads process applies to the external part of the analysis process in which the real world is converted into an idealised finite element model.

The initial stage assumes that the problem has been fully defined in such a manner that the nature of the real world problem is prescribed as outlined in the previous section. As we saw, this is linked with the early stages of the qualification process in which the problem being analysed is specified. The initial qualification stage is described in other consortium documents but one such datum has been established this becomes the input to the first stage of the threads analysis as shown. Thus, the first step i.e. Step 1 requires that the major load paths and loading actions are identified and defined. Although care is taken in assessing the Real World Problem it should be noted that the build-up of knowledge which occurs at later stages in the threads analysis may require changes to the assumed real world definition. This could occur at Steps 5 and 6 where the assumed real world behaviour is re-examined and, possibly, checked against a verification experiment.

The threads analysis envisages a regular process of decomposition of the finite element analysis process with the outputs from each step feeding into one or more subsequent steps. The full process is outlined in figure 1 where the interaction between the various steps is shown. The figure clearly indicates that the decomposition process is not linear and that it can be iterative.

From the error control viewpoint Step 5 is the most important stage being the point at which the viability of the modelling process is assessed. The errors associated with the process must be quantified in order that an error bound or confidence limit is ascribed to the process being decomposed. If this cannot be done then some form of test must be performed. The nature of the test and the parameters to be measured are found in the outputs from Step 3. It is up to the modelling engineer who knows what assumptions are being made to define the test requirement. If any of the assumptions made during any of the Steps 1 to 6 are found to be incorrect as a result of these tests, then the process must loop back to the specified step.

It is, therefore, clear that the threads decomposition links the error analysis process with the qualification process, particularly at steps 1 and 2. However, the role of Features and Primitives begin to play a part in the decomposition process at steps 3 and 4. Their impact on the role and position of testing in the error control process is seen in Step 6 where errors in ascribing a specific behaviour to a Feature/Primitive should be detected and the associated error eliminated.

Although the threads description is applied in the present report to the external consistency problem i.e. the idealisation problem, it can be applied more generally. Applying it to the full analysis process is not covered since it is considered unnecessary as the threads description applied to the modelling process is, in essence, an algorithm which equally applies to all aspects of the analysis procedure. Thus, the same process is repeated at various points within the internal part of the analysis proces when, for example, the elements are selected to match the structural definition from Steps 3 and 4 in figure 1.

FINITE ELEMENT MODELLING THREADS

STAGE 1:	STRUCTURAL ASSESSMENT

STEP 1:	Definition of Load Paths and Loading Actions
	INPUT: Description of real world problem.
	PROCESS: 1. Define major and minor load paths.
	2. Assess loading actions.
	OUTPUT: 1. Description of loading actions and severity.
	2. Partition of primary and secondary structure.

STEP 2:	Definition of Structural Responses
	INPUT: Load Paths
	PROCESS: Assess the structural behaviour of the structure in terms of responses to applied loads.
	OUTPUT: Division of structure into components which correspond to
	- linear static behaviour
	- Linear dynamic behaviour
	- nonlinear static behaviour
	- nonlinear dynamic behaviour

STEP 3:	Breakdown in Detail and Global Analysis	
	INPUT:	Load Paths, Loading Actions.
	PROCESS:	Breakdown of structure into those features which have a relatively smooth action and those with high granularity
		- for high granularity assess if these are to be treated specifically as sub-analyses.
		- if not sub-analyses select appropriate approximate analysis procedure i.e. smearing for plates with stringers.
	OUTPUT:	1. Partition into main and sub-features.
		2. For sub-features define appropriate analysis process.

STEP 4:	Definition of Structural Action	
	INPUT:	1. Outputs from Step 2.
		2. Outputs from Step 3.
	PROCESS:	Define the structural actions of the features defined in process 3 when subject to the loading actions defined in process 2: Thus categorise structural features as those:
		- behaving as membranes, bars, plates, shells, 3-D continuum.
	OUTPUT:	For each feature there is now an associated structural behaviour or behaviours.

STEP 5:	ASSESSMENT	
	INPUT:	1. Outputs from Step 4
		2. Outputs from Step 2
		3. Outputs from Step 1
	PROCESS:	Assess the structural behaviour as described in steps 2 and 4 as an adequate representation of the overall real world problem. If the confidence limit is too low define corroborative tests either at the detailed or global level.
	OUTPUT:	Corroborative tests.

STEP 6:	TEST PROGRAMME	
	INPUT:	Outputs from Step 3.
	PROCESS:	Perform corroborative tests and compare the outputs against the assumed structural actions and responses. If these are inappropriate define changes to response or action assumptions.
	OUTPUT:	1. If tests confirm assumption there is a "proceed" action.
		2. If tests indicate modifications then define changes to assumptions.

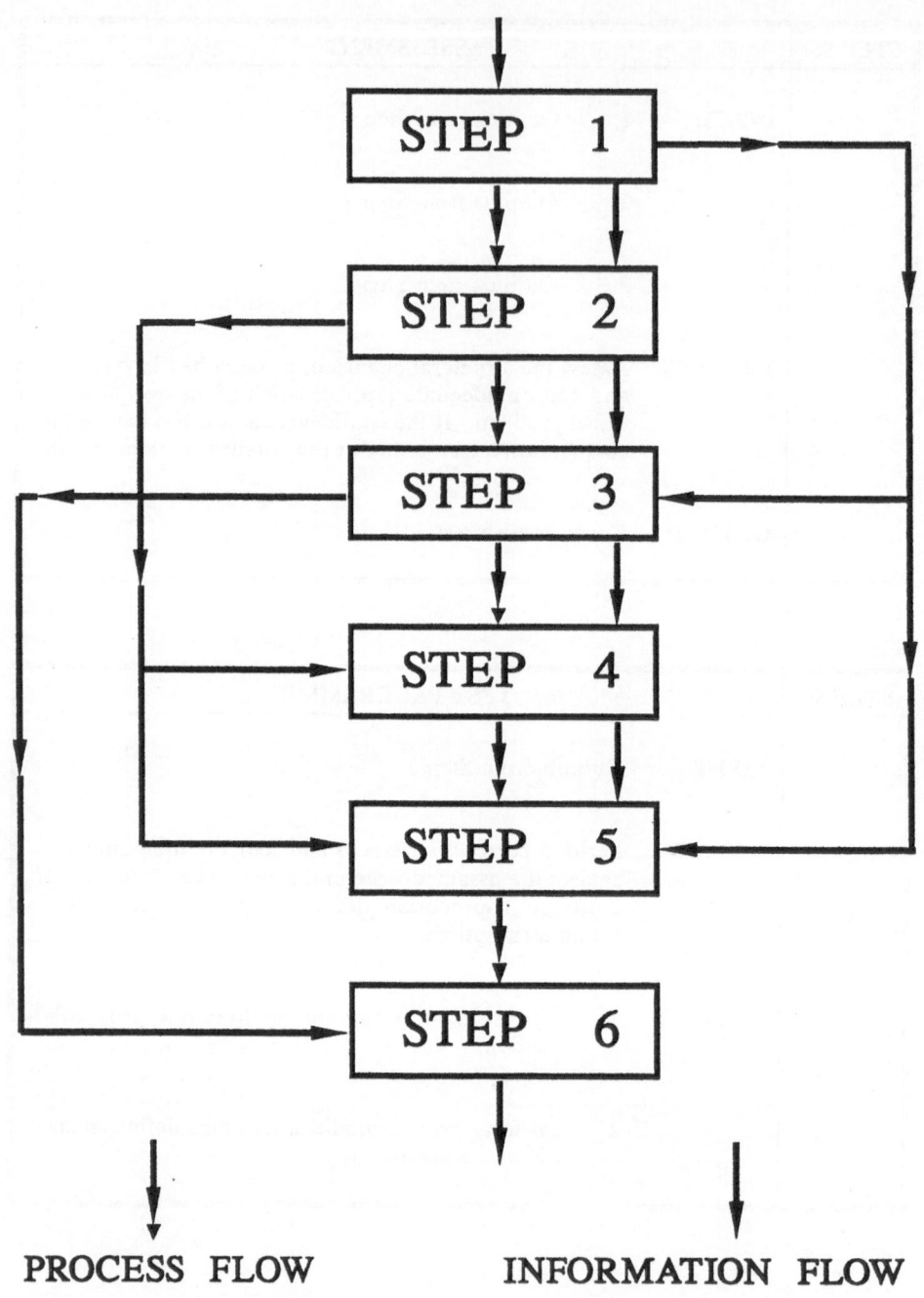

PROCESS FLOW

INFORMATION FLOW

FIGURE 1. INFORMATION AND DECISION FLOWS

3.3 Threads Step 3/4 Features and Primitives

The process of decomposing the analysis process into sub-analyses, structural actions etc., requires a method for defining the appropriate structural components, behaviour etc. In order to handle these aspects the concepts of Features and Primitives is introduced.

The role of Features and Primitives is to decompose the structural description of the analysis problem into structural objects with properties. Thus an Object Oriented description of the structural definition is sought. The Feature is a object which can be treated as a single major item for analysis purposes. Thus the partitioning identified as Step 3 in the modelling threads analysis is the point at which the main analysable components with an identifiable structural behaviour are categorised into Features. In the case of an aircraft, for example, a wing may be identified as a Feature from a stiffness viewpoint since it has a primary role of carrying the aerodynamic loads in bending. A Primitive is a further subdivision into irreducible components associated, possibly, with a specific element or element type. In the case of the wing a specific joint which is to be modelled as a beam may be a Primitive.

For the analysis modelling process the Features and Primitives have errors associated with them as part of their Object description. In this way the errors categorised in Section 4 can be attached to a given piece of structure, finite element, mesh etc. linked to the Primitive.

We may now turn to a formal definition of these two concepts.

A **Feature** is an entity comprising a set of objects in a structural design domain which has a specific characteristic described by it's members, physical condition, function, behaviour, and external action. A structural Feature can be a substructure or components of the structure, or a set of characteristics of a component which exhibit the same behaviour. The members of a Feature clearly result from a decomposition of the Feature into its components.

The physical condition of a Feature comprises a description in terms of such parameters as the geometry, constraints, etc. The behaviour of the Feature describes the performance, the responses to external actions, or the characteristics of its own actions. The function is the objectivity of the Feature to perform in the integrated structure under the influences of its environment or external actions.

A Primitive is the smallest entity in a Feature which is irreducible and does not have a member but only a description and characteristics. The categorization of an entity to become a primitive is based on the possibility of errors emerging from the entity. Narrowing the scope of concern to the smallest entity is meant to confine the source of error to the smallest basic characteristic in such a way that the error is boundable and distinguished from the other sources of error.

A Feature can contain one or several Primitives. Some Primitives could be overlapping in the sense of geometrical or physical meaning. Take, for example, the case of a reinforced hole in the middle of a plate panel: the reinforced hole itself can be considered as a Primitive for a local analysis whilst the whole plate and it's boundaries may be the Primitive when the plate is modelled without a hole in an idealized equivalent condition. This shows that modelling aspects are involved in determining the Features and Primitives.

The following represents some of the characteristics which describe a Primitive:

- A Primitive is an irreducible entity in a Feature that still can be analyzed, focusing on error bounding in the structural analysis.
- A Primitive should have uniform characteristics or attributes, or at least have characteristics which are distributed in a predictable way. For instance the crossectional area of a beam should be uniform or distributed in a regular way along the beam, otherwise the beam should be divided into several predictable elements.
- A Primitive is part of a Feature that needs "special attention" in the analysis, a distinct characteristic or a unique concept, such as the local effective areas, the stress raiser details, the junctions/intersections, the layers in a composite material.

4 Error Taxonomy Related to FE Process

4.1 Broad Categorisation:

The taxonometric approach has proved very effective in providing a framework for error control in the development of software systems and has been adopted, in SAFESA, as a basic methodology for the application of F.E. systems in the solution of real world problems. Thus, in order to evaluate Finite Element Analysis, an effective classification of FE errors is required. There exist numerous sources of potential errors in the Finite

Element solution of design problems. A broad classification of such errors is presented by Morris [1]. In what follows, an attempt is made to expand on error classification as found in the latter reference, taking into account NAFEMS guidelines to FE practice [4], remarks and suggestions made by SAFESA partners (especially those made by Nuclear Electric, Fox [3]) and the work of Beizer [2].

Errors are broadly classified into four main categories (see figure 2):

System Software Errors	CLASS 0
Formulation Errors	CLASS 1
Analysis Errors	CLASS 2
Modelling Errors	CLASS 3

CLASS0 - System Software Errors:
This category would include all types of errors which are not specifically peculiar to FE but are rather often encountered when using any computerised system. Errors under this class would cover any failure caused by a piece of software used during (or developed for) the FE process. Virtually, all of the errors that can occur in software may be considered *human errors* of one type or another. However, they can be classified under five distinct categories: software design errors, implementation errors, operator errors, maintenance errors, and testing errors.

0.1 Software Design Errors:
The design phase is the area in which introduced errors will have the greatest impact. Errors of this category will be propagated throughout the design to the final output of the software. Once detected, these errors are difficult and very expensive to fix. It is believed that a design error discovered during the execution will cost 20 times the amount to correct had it been discovered during the design phase.

123

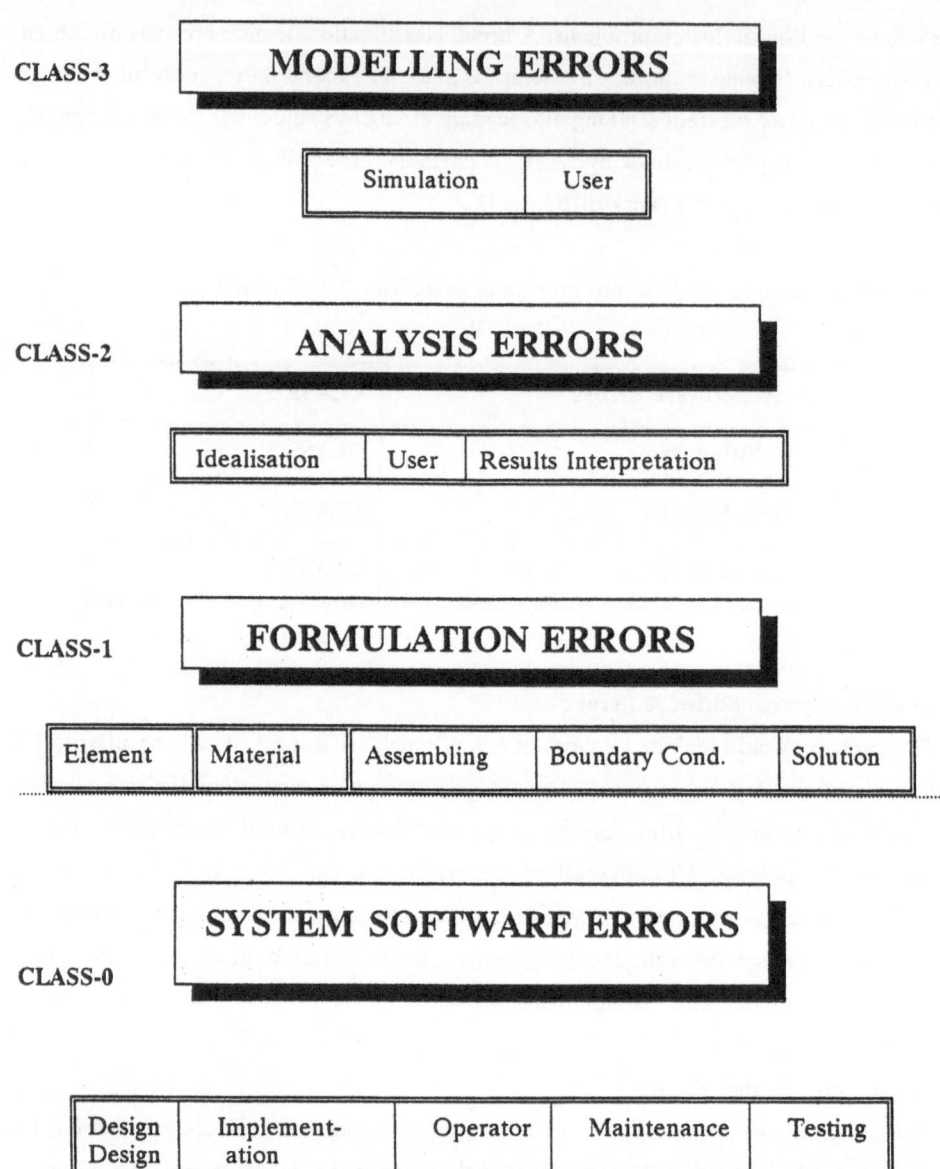

CLASS-3 — MODELLING ERRORS

Simulation	User

CLASS-2 — ANALYSIS ERRORS

Idealisation	User	Results Interpretation

CLASS-1 — FORMULATION ERRORS

Element	Material	Assembling	Boundary Cond.	Solution

CLASS-0 — SYSTEM SOFTWARE ERRORS

Design Design	Implement-ation	Operator	Maintenance	Testing

FIGURE 2: A BROAD CATEGORISATION OF F.E. ERRORS

0.2 Implementation Errors:

This category includes all types of programming errors such as lexical, syntactical, logical, ...etc. This category may also include bugs due to the changes between the different hardware platforms on which the software (assumed correct) is running.

0.3 Operational Use Errors:

A wide variety of human errors can occur during the use of the software. Typical examples of this type of errors include misunderstood prompt, misunterpretation of documentation, ...etc. Preventing these errors requires the involvement of Human Factors in the early design and analysis of the system software. This includes the development of design requirements, recommendation and guidelines, particularly with respect to the user interface.

0.4 Maintenance Errors:

The errors introduced during software maintenance closely follow the types of errors made during the development. However, the errors are often aggravated, in both number and complexity, by a lack of familiarity with the overall design of the system and the complex interactions that occur in it. One study concluded that for every software error corrected during maintenance, another was either introduced or uncovered. A clear and comprehensive documentation of the software design structure would help in preventing this type of error.

0.5 Testing Errors:

This category includes errors that occur as a result of the technique used for verifying the software. The way in which the function check is implemented in the code may negate the effectiveness of that check (i.e. not meet the intent). Similarly, a modification in a test procedure may result in a failure to test a given path of the software. Such errors affect the output of the verification process and have to be dealt with.

Note, however, that little attention will be given to this class of errors within the scope of SAFESA project. It is assumed that the software used for the FE analysis (e.g. Bersafe, Asas, Nastran,...) has already been verified and validated beforehand. The only remaining preoccupation, as far as this class of errors is concerned, is to deal with errors related to any application software that has been developed for the purpose of the FE analysis.

CLASS 1 - Formulation Errors:

Definition of Class 1 errors is based on the assumption that the FE software is error-free, i.e. errors generated in software development were detected and corrected during testing and verification phase. This category would include all types of errors generated by running the error free software to create a 'Numerical Solution' from the 'Solution Model'.

Excluding the errors that are due purely to the operator inexperience or carelessness, this category would include the following types of errors:

1.1 Element Errors:

Errors due to element formulations (discrepancy between an element formulation and related theory).

1.2 Material Errors:

Errors due to formulation of constitutive equations.

1.3 Assembling Errors:

Errors due to inadequate integration of the FE model.

1.4 Boundary Condition Errors:

Errors related to the treatment of the natural and geometrical boundary conditions and contact algorithm (constrains).

1.5 Solution Errors:

Errors related to the techniques employed to achieve a solution of the series of equations describing behaviour of the FE model.

CLASS 2- Analysis Errors:

This class is concerned with errors which occur because the finite element cannot match the underlying theory. Most of these errors typically occur during the steps from the 'Geometrical Model' to the 'Solution Model'. Included in this category are the following types of errors:

2.1 Idealisation/discretisation:

.Error in selecting the analysis type

.Wrong Idealisation type

.Error in the choice of the solution model

.Wrong element type

.Inappropriate material model

.Inappropriate load model

.Etc...

2.2 User Errors:

2.2.1 Data Errors: Bugs in the definition, structure or use of data.

.Use of elements outside of legitimate range

.Incorrect boundary conditions

.Incorrect loads

.Improper reading of data files

.Inability to interpret results properly

.Etc...

2.2.2 Unfamiliarity with the pre- and post-processors:

.Misunderstanding a prompt message

.Misinterpretation of a diagnostic

.Ignoring the built-in checks

.Etc...

2.2.3 Negligence or carelessness:

.Previous miscorrection of an error

.Other negligence errors

2.3 Results Interpretation:

Errors that occur due to a wrong interpretation of results.

CLASS 3 - Modelling Errors:

Included in this class are both 'Simulation' and 'User' errors as defined in [1]. Simulation errors are generated during the process of passing from the designed structure to the geometrical model, yielding an inadequate representation of the real world structure (for instance solid or wire frame model). User errors would include any bug that was inserted in the system as a consequence of the user's inexperience or carelessness during this phase of the FE analysis.

3.1 Simulation Errors:

3.1.1 Error in the Definition of the real world problem.

    .Load type

    .Fabrication type

    .Material type

    .Etc...

3.1.2 Error in the Definition of Major Structural Components to be analysed.

    .Load path

    .Geometry

    .Fabrication type

    .Material type

    .Etc...

3.1.3 Error in the Definition of Structural Primitives for each Component.

    .Structure type

    .Joint type

    .Intersections

    .Etc...

3.2 User Errors:

Errors in this category are quite similar, in nature, to the ones outlined in Class 2.2 above.

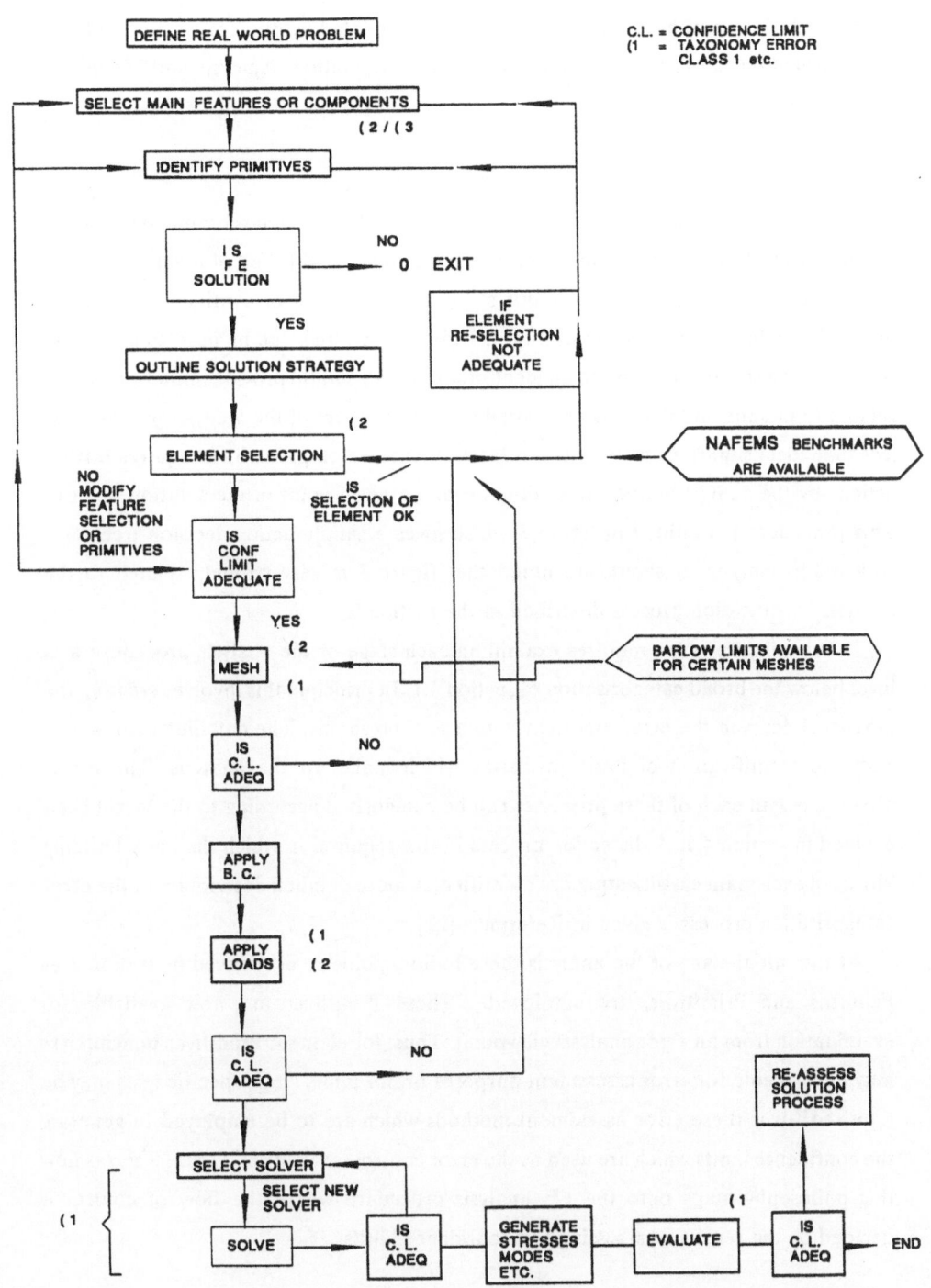

C.L. = CONFIDENCE LIMIT
(1 = TAXONOMY ERROR CLASS 1 etc.

DEFINE REAL WORLD PROBLEM

SELECT MAIN FEATURES OR COMPONENTS

(2 / (3

IDENTIFY PRIMITIVES

IS F E SOLUTION

NO

0   EXIT

YES

IF ELEMENT RE-SELECTION NOT ADEQUATE

OUTLINE SOLUTION STRATEGY

(2

ELEMENT SELECTION

NAFEMS BENCHMARKS ARE AVAILABLE

NO MODIFY FEATURE SELECTION OR PRIMITIVES

IS SELECTION OF ELEMENT OK

IS CONF LIMIT ADEQUATE

YES   (2

MESH   (1

BARLOW LIMITS AVAILABLE FOR CERTAIN MESHES

IS C. L. ADEQ

NO

APPLY B. C.

APPLY LOADS   (1 (2

IS C. L. ADEQ

NO

RE-ASSESS SOLUTION PROCESS

SELECT SOLVER

SELECT NEW SOLVER

(1

SOLVE

IS C. L. ADEQ

GENERATE STRESSES MODES ETC.

EVALUATE   (1

IS C. L. ADEQ

END

Figure 3. Tree Chart for a Typical F.E. Analysis

129

It is worthwhile emphasising that for SAFESA' related work more effort will be concentrated on the last three classes of errors, Formulation, Analysis and Modelling errors.

4.2 Detailed Taxonomy

The process of classifying the errors in the total FE analysis must be continued at a more detailed level if an effective control procedure is to be established in a genuine safety critical environment. The target at this stage is the creation of confidence factors or limits. These limits allow the analyst to provide values which can be used to assess the results from a specific analysis. In an ideal situation they should provide numerical values related to bounds on the errors associated with each aspect of the analysis process. For less than ideal situations they should at least provide a stop gate which requires further action by the analyst before proceeding beyond a given point in the solution process. This procedure is outlined in figure 3 which gives a simple action/decision tree for a typical FE analysis. It should be noted that figure 3 is very strongly related to the analysis classification process described in the section 3.

The detailing process requires examining each stage of the analysis procedure at a level below the broad categorisation of section 4.1. In principle this involves creating the next level down in the error hierarchy with a new taxonomy. The detailing commences with the identification of basic processes which make up the analysis. The errors associated with each of these processes can be categorised according to the broad basis decided in section 4.1. A dissection process is now required in which the basic building blocks of each main classification are identified. A more detailed description of the error categorisation process is given in Reference [21].

At the initial stage of the analysis these building blocks, introduced in section 3 as Features and Primitives, are employed. These Primitives are now available for examination from an error analysis viewpoint. Thus, for element Primitives benchmarks may be available for error assessment purposes or, for joints (say), specific tests may be required. It is these error assessment methods which are to be employed to generate the confidence limits which are used as the error control mechanism. Figure 3 shows how this philosophy maps onto the FE analysis procedure where the flow of control is decided by the results of assessing the confidence limits.

REFERENCES:

[1] Morris, A.J. " An Approach to the Validation of Finite Element Codes ", Internal Report, College of Aeronautics, Cranfield Institute of Technology, UK.

[2] Beizer B. " Software Testing Techniques ", Van Nostrand Reinhold Publishers, New York, 1990.

[3] Fox, M.J.H. "Comments on Software Error Classification", SAFESA Technical Note 9034/TN/NE/0001/16.3.92

[4] Schneidewind N.F. & Hoffmann H.M. " An Experiment in Software Error Data Collection and Analysis ", IEEE Trans. on Software Engineering, Vol. SE-5, No. 3, May 1979, pp 276-286.

[6] Endres A. " An Analysis of Errors and Their Causes in System Programs", IEEE Trans on Software Engineering, Vol. SE-1, No. 2, June 1975, pp 140-149

[7] Nakajo T. & Kume H. " A case History Analysis of Software Error Cause Effect Relationships ", IEEE Trans on Software Engineering, Vol 17, No. 8, August 1991, pp 830-838

[8] Zienkiewicz, O.C; "The Finite Element Method", 4th Ed., vol 1, Mc Graw-Hill Book Co., London, UK, 1989

[9] Tong, P. & Rossettos, J.N. "Finite Element Method", The MIT Press, Cambridge, 1977

[10] Norrie, D.H & Devries, G "An Introduction to Finite Element Analysis", Academic Press, New York, 1978

[11] Weiskamp, K, et al; "Object Oriented Programming with Turbo C++", John Wiley & Sons Inc., New York, 1991

[12] Hughes, T.J.R "The Finite Element Method", Prentice Hall International, Inc., New Jersey, 1987

[13] Babuska, I. "Accuracy Estimates and Adaptive Refinements in Finite Element Computations", John Wiley & Sons, New York, 1986

[14] Walker, M.G. "Managing Software Reliability", North Holland, New York, 1981

[15] Myers, G.J. "The Art of Software Testing", John Wiley & Sons, New York, 1979

[16] Huebner, K.H. "The Finite Element Method for Engineers", John Wiley & Sons, New York, 1975

[17] Cook, R.D. "Concepts and Applications of Finite Element Analysis", John Wiley & Sons, New York, 1989

[18] Deutsch, K.W. "On Theories, Taxonomies, and Models as Communication Codes for Organising Information", Behavioral Science 11, pp1-17. 1966

[19] Barlow, J. "Critical Tests for Element Shape Sensitivity", Proc. 6th World Congress on Finite Element Method Banoff Canada pub, J Robinson & Asociates, 1991

[20] NAFEMS "The Standard NAFEMS Benchmark", NAFEMS (National Agency for Finite Element Method & Standard), DTI (Department of Trade & Industry), National Engineering Laboratory, Glasgow, Oct, 1990

[21] Hadi, A.K. "T91 Composite Elevator Design", MSc Design Project Thesis, Aerospace Vehicle Design, College of Aeronautics, Cranfield Institute of Technology, Cranfield, May 1992

[21] Hadi, A.K. "Qualification Process", internal draft paper, College of Aeronautics, Cranfield Institute of Technology, Cranfield, June 1992

[22] Standards and design manual "BCAR" and "JAR" of Civil Aviation Authority London, and military standard "Defence Standard 00-970", ESDU data sheets

[23] NAFEMS "Guidelines to Finite Element Practice", NAFEMS (National Agency for Finite Element Method & Standard), DTI (Department of Trade & Industry), National Engineering Laboratotry, Glasgow, 1986

[24] NAFEMS "A Finite Element Primer", NAFEMS (National Agency for Finite Element Method & Standard), DTI (Department of Trade & Industry), National Engineering Laboratotry, Glasgow, 1986

[25] Howard, H.C. et al "A Primitive-Composite Approach for Structural Data Modelling", ASCE Journal for Computing in Civil Engineering, Special Issue on Databases, 1991

[26] MSC "MSC Nastran Manual, version 66A",

[27] MSC "MSC-XL Manual", graphical post-processor for Nastran

Using the Functional Programming Language Haskell to Specify Image Analysis Systems *

Ian Poole and Derek Charleston

MRC Human Genetics Unit
Crewe Road
Edinburgh EH6 4AT

Brian Finnie

The Centre for Software Engineering Ltd.
Bellwin Drive
Flixborough, South Humberside DN15 8SN

Abstract

This paper demonstrates the feasibility (and discusses the advantages) of using a functional programming language as a vehicle for both formal specification and final implementation of diagnostic imaging systems. We show how the syntax of a particular language, Haskell, can be used to record implicit specifications, general purpose theorems and proofs. An example of the derivation of a constructive definition from an imperative one, notated in Haskell is given. Some practical issues concerning the implementation of imaging systems in a functional language are discussed briefly.

1 Introduction

1.1 Background

Diagnostic imaging systems developed for the medical sector invariably involve large, complex and subtle controlling software. Examples are semi-automated systems for chromosome karyotyping and cervical cytology screening. Such systems involve the real-time control of a microscope and camera, image analysis, pattern recognition

---

*This work is partly supported by an ITD grant under the Safety Critical Systems programme. The project title is "SADLI: Safety assurance in diagnostic laboratory imaging"; collaborators are The Centre for Software Engineering Ltd., MRC Human Genetics Unit and Cambridge Consultants Ltd.

and operator interaction. Being in the medical domain, there is increasing concern for the "correctness" of these systems, especially when used in mass screening programmes. In this context we are investigating the applicability of safety-critical methodologies to diagnostic imaging systems. In this paper we focus on the use of *formal methods* of software development in this application domain, and advance the view that a *functional programming language* can be tailored to serve as a flexible tool for both system specification and execution. The reader unfamiliar with functional programming is encouraged to study the brief introduction which is given as an appendix before proceeding.

1.2 Formal specification

Software production is essentially the task of proceeding from some (often imprecise) statement of *requirements* to a form which is sufficiently precise to execute on a computer; from the hopelessly *in*formal to the perfectly *formal*.

Before proceeding we should perhaps clarify our intended meaning of the word "formal". We say that a language (or notation) is formal if sentences of that language can be ascribed a precise mathematical meaning. Thus both Pascal and Z [1] are formal languages, but English is not. The simplicity of the semantics, influencing the ease with which one can reason in the language, is a separate issue which we discuss below.

Figure 1 shows a much simplied view of the "traditional" and "formal" methodologies of software production.

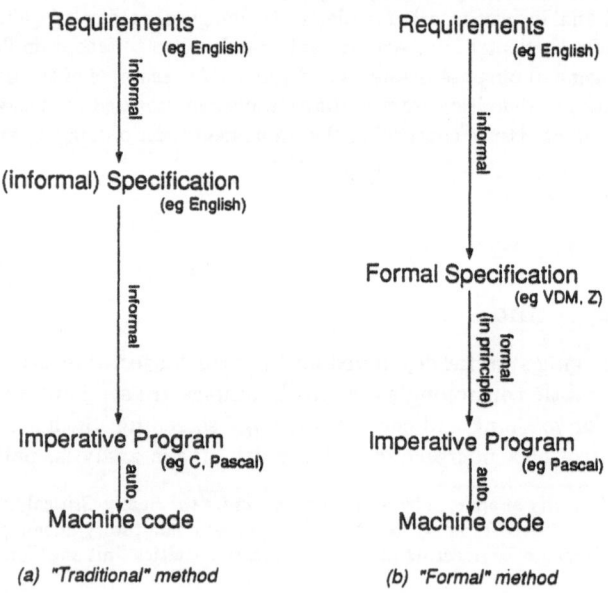

Figure 1: Traditional and formal methods of software production (simplified).

The traditional method (a) interposes an (informal) *specification* document, using natural language and diagrams, intended to state in detail the functionality of the system to be constructed. This forms the basis for the production of the software in an appropriate high-level language. (The intervening design documents or iteration steps are not shown.) Under this traditional method, an entirely formal specification of the system is not achieved until nearly the end of the process, and this is in the form of an imperative programming language (eg, C, Ada, Pascal ...) which is verbose and extremely difficult to read and reason about.

The formal method (b) demands that the specification document be written in a "formal specification language" — examples of such language are Z and VDM[8] resulting in a so-called *formal specification*. As well as having formal and therefore unambiguous semantics, these languages have a strong mathematical flavour, permitting succinct specifications which are tractable to formal reasoning. They allow a relational style of specification (also termed *implicit* specifications), which is often clearer and shorter than a constructive form. Note that implicit specifications can be deliberately ambiguous; for example we might implicitly define y by $y^2 = x$, permitting a positive or negative value of y for any (positive) value of x.

The task of obtaining an executable imperative program from a formal specification is a largely manual one, although it is possible (with difficulty, see below) formally to prove that the program is consistent with the specification.

1.3 Problems of separating specification and program

Although some developments using the above approach have been successful, their use is not widespread and we have identified three particular difficulties.

Firstly, support tools remain difficult (or expensive) to come by. As a minimum, one needs utilities to check for correct syntax, type consistency and completeness of specifications. In the case of Z, which uses a wide range of non ascii characters, special input and formatting tools are also required. Without such tools, the formal specification will be a paper (or word-processor) document only, and will inevitably contain many typographic (or deeper) errors. There is also the danger that the specifier will use the language loosely and slide into informal pseudo-code.

Secondly, the need to translate from one formal language (eg VDM) into another formal language (eg Pascal), we consider an unnecessary overhead. Hitherto this has been a largely manual process. Whilst it is possible in principle formally to prove the correspondence of the imperative code with the original specification (providing a formal semantics for the target language is available), the opaque semantics of imperative languages (i.e. side-effects) make this difficult for all but trivial systems. Work aimed at *automatically* translating an executable subset of a specification language (eg from VDM to Pascal [11]) is indeed interesting, and we would suggest that the selected sub-set is effectively a functional programming language, resulting in a convergence with our own ideas.

Finally, it requires considerable discipline to ensure that formal specification and executable program remain consistent through subsequent modifications.

1.4 Outline of our methodology

It is our view that the distinction between (formal) specification and executable program is overstated. As mentioned above, we consider the crux of the software engineering task to be the transition from the informal to the formal. Whatever one's choice of formal language (Pascal, Ada, Haskell (see appendix), VDM ...) this transition will always be time-consuming and intellectually demanding, requiring expert knowledge of the application domain. In our view, the transition of a formal specification into executable code is a lesser problem and should not require the use of a separate language.

We therefore desire a single language with the following properties:

Readability — at least to an expert in the application domain, with a limited knowledge of the formal language, enabling them to reason as to the propriety of the specification with respect to the original requirements.

Mathematically based — permitting succinct specifications in both implicit and constructive forms, possessing semantics which are tractable to formal reasoning.

Efficiently executable — to serve as the final implementation language.

Note however that we do not expect all specifications within the language to be executable — implicit definitions are in principle non-executable (for example, there may not be a unique solution) so we accept that these must first be transformed (within the language) into constructive forms.

In the remainder of this paper we will demonstrate that a lazy functional language such as Haskell [1] can be tailored to fulfill these requirements.

For those not familiar with functional programming languages, a brief introduction and further reading list is given as an appendix to this paper.

Figure 2 summarises our proposed methodology. The formal specification is written in Haskell, with the use of implicit definitions where convenient. Examples of this are given in section 3. The specification document includes descriptive text and diagrams where helpful (see section 2). This first form of the specification will most likely be non-executable, but it will never-the-less be checked by the compiler for syntax and type consistency, and for completeness. undefined).

The next step is to provide constructive definitions for any implicitly defined functions. However, the original definitions remain in place, and continue to be checked for type consistency, since these most clearly document the functionality of the system. We should make it clear from the outset, however, that the Haskell language system *cannot* verify the *logical* accuracy of these transformations — that would require a semi-automated theorem proving system such as "Starship" [14]. This is an interesting possibility, but one we have not yet investigated. In safety critical areas of the system, formal proofs should be given that the constructive definitions satisfy the original specification, and as will be seen, these too may be

---

[1] It is worth pointing out that we have selected Haskell mainly because interpreters and compilers are freely available for the language; Haskell is closely related to Miranda [16] (trademark of Research Software Ltd.), and at least for the issues addressed in this paper, Miranda would suffice perfectly.

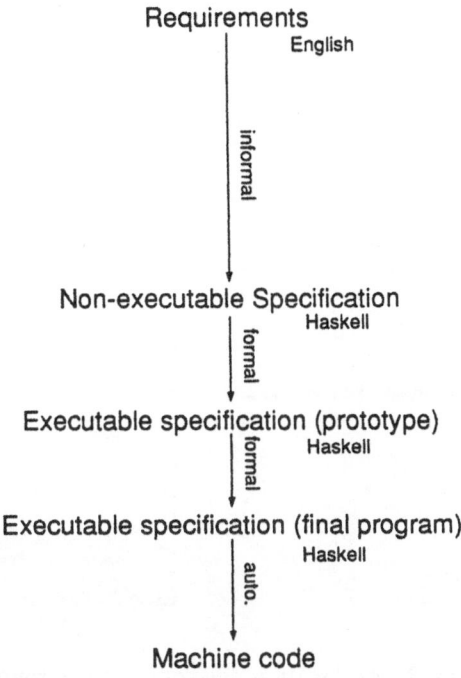

Requirements
English

informal

Non-executable Specification
Haskell

formal

Executable specification (prototype)
Haskell

formal

Executable specification (final program)
Haskell

auto.

Machine code

Figure 2: Haskell as a specification and implementation language.

presented in checkable Haskell syntax. This will result in an executable specification which can be tested as a prototype.

Finally, further transformations are carried out to improve execution properties, again providing formal proofs that the properties of the original specification have been conserved. After compilation, this delivers the final program.

In the following three sections we show how a Haskell script can fulfill the three requirements set out above.

2 Legibility of Haskell scripts

Haskell supports *literate scripts*, sometimes known as "inverted comments". The normal convention of indicating comments by a special character sequence is reversed, so that all text is treated as comment by default, and formal code must be specially flagged. In a literate Haskell script, code is indicated by a '>' character at the beginning of the line, thus:

```
> eg_hello = "This is a formal greeting"
```

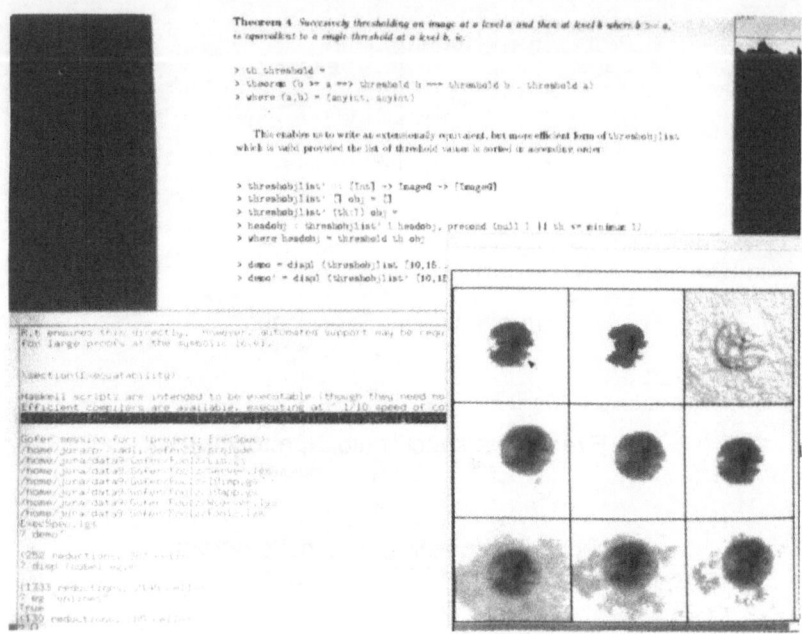

Figure 3: Workstation development environment

This seemingly trivial facility turns out to be very powerful when combined with an ascii based typesetting system such as LaTeX [10]. The LaTeX source for this document is a compilable Haskell script, and the constructively defined values, such as the one above, can be evaluated from the interpreter, whilst the printed form of the document includes imported diagrams and is (we hope) attractively formatted. Facilities such as the automatic production of contents-tables and numbering of theorems, etc. can assist in producing a more easily referable specification document. Given that Haskell syntax is terse — specifications, even when written constructively, are typically between 1/4 to 1/10 the length of their imperative counterparts — the volume of explanatory text and diagrams usually outweighs the formal code. When complete, the source document *is* the program.

Standard previewing tools enable the typeset document to be viewed at a workstation. Figure 3 is a photograph of the screen taken whilst this paper was being developed. The left hand window shows the functional language interpreter [2] linked with the editor open on the source file. Centre and partly covered is the GNU "Ghostview" previewer (courtesy of the Free Software Foundation) showing the typeset document. On the right is an image display, being used to test the functions `threshobjlist'` and `sobel` which are defined in section 4.1.

---

[2]In fact this is an interpreter for the language Gofer[9], a close dialect of Haskell. As far as the facilities used in this document are concerned it is identical to Haskell, except that Haskell requires a "module" declaration which Gofer does not support. We use the Gofer interpreter because it is friendlier and faster to use than any we have (yet) found for Haskell.

3 Mathematically based specifications in Haskell

In this section we show in detail how Haskell can be used to record succinct, mathematically styled specifications and general purpose theorems. In particular we show how implicit definitions can be mixed freely with constructive forms (for which the language is primarily intended).

3.1 Infra-structure

To achieve our aims we must first define some "meta functions" — these would be included as an appendix to a specification document. It is indicative of the flexibility and extensibility of Haskell that these facilities can be provided *within* the language.

Some of the functions we introduce are inherently non-executable, and so will be "defined" in terms of non_exec.

```
> non_exec msg = error msg
```

The standard equality operator == is, for obvious reasons, not defined for general functions. In non-executable specification we will wish to notate equality between pairs of *any* type, including functions, and so we define a new operator === as follows.

```
> infix 4 ===
> (===) :: a -> a -> Bool
> f === g = non_exec "Equality of functions cannot be evaluated"
```

The infix 4 declaration makes === non-associative and specifies its binding priority. This will keep the compiler happy, whilst displaying the error message if we ever try to evaluate it.

An "implication" operator will be useful when stating theorems and constructing proofs. The => symbol is already used in Gofer, so we will use ==>. The infixl 3 declaration makes the operator left associative with a binding priority just below that of ===.

```
> infixl 3 ==>
> (==>) :: Bool -> Bool -> Bool
> x ==> y = if x then y
>                else error "==> is strictly true, but premise is false!"
```

Implicit definitions are presented as shown in the following example:

```
> sqrt :: Int -> Int
> sqrt x = implicit (
```

```
>                      y,
>                      precond (x >= 0) && postcond (y * y == x)
>         )
>                      where y = notyet
```

'precond' and 'postcond' stand for the 'Pre-condition' and 'Post-condition', familiar in VDM. Both are simply defined as Bool functions and so are connected by the logical *and* operator '&&'. [3]

```
> precond True = True
> precond False = error "Pre-condition false"

> postcond True = True
> postcond False = error "Post-condition false"

> implicit (val, cond)
>       = if cond then val
>                  else error "Implicit definition fails"
```

The following will also prove useful:

```
> notyet = non_exec "Definition not yet available"
> universal = non_exec "Cannot evaluate 'universal'"
> universalInt :: Int
> universalInt = universal
```

We use universal to indicate universal quantification [4].

A *derivation* is a tuple of the derived value (probably a function) and a (hopefully true) boolean expression representing the proof of the derivation. Two functions are provided to extract the result and the proof from the derivation.

```
> type Derivation a = (a, Bool)
> result_of (result, proof)  =  if proof then result
>                                          else error "Proof is invalid"
> proof_for (result, proof) = proof
```

---

[3]Actually, we consider the terms *pre-condition* and *post-condition* to be unfortunate — they are suggestive of imperative thinking. Really one just wishes to state ("**assert**" perhaps) the required property of the function as a single boolean expression. The *pre-condition* is really just a restriction on the domain of the function — a predicate which does not involve the function its self.

[4]This is rather glib. We have work to do in providing an exact semantics for expressions involving **universal**. Similarly, we need to define more precisely the intended meaning of === when used to equate functions.

A theorem is recorded as a (hopefully true) Boolean expression.

```
> theorem True = True
> theorem False = error "Theorem is false"
```

3.2 Some useful theorems and definitions

To demonstrate the above infra-structure and to prepare the ground for a more substantial example, consider the following.

Theorem 1 (map) *The composition of two function maps is equivalent to the map of the composition of the two functions, ie:*[5]

```
> th_map = theorem (map f . map g === map (f . g))
>          where (f,g) = universal
```

This is nothing more than a formal statement of the obvious. It is straight forward, if a little tedious, to provide a rigorous inductive proof, beginning from the definition of map in Haskell's Standard Prelude.

The standard functions lines :: String ->[String][6] divides a newline (\n) formatted string into a list of strings, one list for each line. The standard function unlines :: [String] ->String joins a list of strings with newlines.

Theorem 2 (unlines) unlines *is the inverse of* lines, *ie* [7]:

```
> th_unlines_lines =
>       theorem (unlines . lines === id)
```

Again, this can be proved from the definitions in the Standard Prelude.

Note that there is no exact inverse of unlines, since it is not a one-to-one function; for example,

```
> eg_unlines = unlines ["Hello", "World"] == unlines ["Hello\nWorld"]
```

evaluates to True, since both sides of the equality evaluate to "Hello\nWorld".

Analogously, words :: String > [String]) divides a string of white-space separated words into a list of strings

---

[5]map applies a function to each element of a list. Function composition is notated by ".", ie (f . g) x === f (g x).

[6]lines :: String ->[String] reads as "lines is a function which maps a String to a list of Strings". A String is a list of characters.

[7]id is the identity function

Theorem 3 (unwords) unwords *is the inverse of* words, *ie:*

```
> th_unwords_words =
>         theorem (unwords . words === id)
```

and

Theorem 4 (strToInt) strToInt *is the inverse of* showint, *ie:*

```
> th_read =
>         theorem (strToInt . showInt === id)
```

3.3   A derivation from a simple implicit definition

Following is the constructive definition for a function to format a list of lists of integers as multiple line text, for display or saving to a file.

```
> writetable :: [[Int]] -> String
> writetable = unlines . map unwords . map (map showInt)
```

Suppose we now require a function to read back the string produced by writetable to recover the original structure, ie,

```
> readtable :: String -> [[Int]]
> readtable = implicit (
>                       result_of read_table_derivation,
>
>                       postcond (readtable . writetable === id)
>              )
```

Whilst writing the specification without concern for executability, the expression result_of read_table_derivation would simply be replaced by implicit

By applying the theorems set out in the previous sub-section, we can formally derive the constructive definition of readtable from the above implicit form. The derivation is presented as a single Boolean expression, composed with the implication operator ==>. Of course, Haskell cannot ensure the truth of the derivation, but does perform valuable syntax and type checking.

```
> read_table_derivation
>   = (readtable', proof)
>   where
```

```
>    proof =
>                -- the implicit definition is:
>
>         readtable . writetable === id
>
>           ==> -- by substitution of writetable ...
>
>         readtable . unlines . map unwords . map (map showInt) === id
>
>             ==> -- by theorem th_read we expand the rhs 'id'
>
>         readtable . unlines . map unwords . map (map showInt) ===
>             map (map (strToInt . showInt))
>
>             ==> -- by th_map (twice) ...
>
>         readtable . unlines . map unwords . map (map showInt) ===
>             map (map strToInt) . map (map showInt)
>
>             ==> -- by canceling 'map (map showInt)' on both side ...
>
>         readtable . unlines . map unwords === map (map strToInt)
>
>             ==> -- via an analogous process, using th_unwords_words ...
>
>         readtable . unlines === map (map strToInt) . map words
>
>             ==> -- similarly via th_unlines_lines we finally get ...
>
>         readtable === readtable'
>
>    readtable' = map (map strToInt) . map words . lines
```

The final line of the derivation is a constructive definition of readtable. Readtable is now an executable function, so that

```
> eg_readtable = readtable "1\n2 3\n4 5 6\n"
```

evaluates to

```
        [[1], [2,3], [4,5,6]] .
```

143

4 Implementation issues

Haskell has been developed primarily as a functional programming language which can be efficiently implemented. Compilers are available, generating code which will execute (only) one order of magnitude slower than the corresponding C version for typical programs, and they are improving all the time. The prospect of automatically mapping onto parallel architectures (the absence of side effects make this inherently easier than for imperative languages) may ultimately turn the efficiency question on its head, though such sytems remain as yet in the research laboratory.

Whilst "typical" functional programs execute in acceptable time and space, we have found they are much less efficient for programs doing pixel level image processing. In a language like C, programs to process large arrays of data (ie images) can be highly optimised, leaving a functional version behind by nearly two orders of magnitude.

We have thus been forced to adopt a pragmatic approach, essentially adopting a pre-existing C coded image processing library to implement the pixel level image processing functions, with a clean interface allowing them to be called from Haskell. Thus the application specific code (the programmer intensive part) is implemented in Haskell, and the pixel crunching (the processor intensive part) is in C. Full details can be found in [13]. The following sub-section gives a brief summary.

4.1 Image analysis applications in Haskell

The abstract data type `ImageG` is introduced to represent an arbitrary shaped region with associated integer grey-scale values. [8]

A large number of basic operations are defined for type `ImageG`, following are two examples.

threshold — given a grey-scale image, return an image whose domain covers only those pixels having a value greater than, or equal to, a given value.

```
threshold :: Int -> ImageG -> ImageG
```

labellist — return a list of the connected sub-components of an image. This function is usually applied after thresholding to achieve segmentation.

```
labellist :: ImageG -> [ImageG]
```

These datatypes and functions are closely modeled on those provided by the C coded image processing library[12], known as "Woolz", and indeed it is this library which is used for their low level implementation. All image data is held in a separate server process — the "Woolz server". The server accepts *evaluation requests* via a Unix socket. Evaluation requests are in the form of null-terminated character strings, the first part of which identifies the function, with parameters following. Images held by the server are identified by a string-encoded integer, the *image-id*.

When the result of an evaluation request is an image, the server holds the image in memory and returns its identity as a string. The server also provides a display

---

[8]This is a simplified description; there are also types to represent binary images, polygons and boundaries, as well as a type *class* **Image** to collect all types which define a spatial *domain*.

request which causes a given image to be displayed in a graphics window, as can be seen in figure 3.

4.2   Some image analysis examples

As part of an algorithm to select the optimum threshold value (see `threshold` above) for a cell nucleus, a function is needed to deliver a list of images, each being the original image thresholded at different values taken from a list, i.e. we require:

```
> threshobjlist :: [Int] -> ImageG -> [ImageG]
> threshobjlist thl obj = [threshold x obj | x <- thl]
```

This is a very clear specification of what is required, but in fact its efficiency can be substantially improved by altering its intensional properties in the light of the following simple theorem.

Theorem 5 (threshold) *Successively thresholding an image at a level* a *followed by level* b *where* b >= a, *is equivalent to a single threshold at level* b, *ie:*

```
> th_threshold =
>         theorem (b >= a ==> threshold b === threshold b . threshold a)
>         where (a,b) = (universalInt, universalInt)
```

This enables us to write an equivalent, but more efficient form of `threshobjlist` which is valid provided the list of threshold values is sorted in ascending order, and to provide a formal justification for the use of `threshobjlist'` by the use of the theorem `th_threshobjlist`.

```
> threshobjlist' :: [Int] -> ImageG -> [ImageG]
> threshobjlist' [] obj = []
> threshobjlist' (th:l) obj =
>         headobj : threshobjlist' l headobj
>         where headobj = threshold th obj
```

This is more efficient since fewer pixels need to be examined at each stage.
    As a further image processing example, here is a simple edge detector:

```
> sobel :: ImageG -> ImageG
> sobel im
>    = vs 'add' hs
>      where
>          vs = (modulus . convolve vsobel) im
```

145

```
>        hs = (modulus . convolve hsobel) im
>        vsobel = [[-1, 0, 1],
>                  [-1, 0, 1],
>                  [-1, 0, 1]]
>        hsobel = transpose vsobel
```

Finally, the function which generated the display in figure 3 is:

```
> demo' = displ (threshobjlist' [10,15..45] egim ++ [sobel egim])

> egim :: ImageG
> egim = (hd . readImageGlist)
> "/home/jura/dataa/pr/sadli/Images/e9107839_1.od"
```

5 Conclusions

We are certainly not the first to have proposed using a functional programming language as a specification and prototyping tool. Work such as that described in [5] and [2] — based on the executable specification language "me too" — have been germane to our own. This paper has advanced these ideas by showing how a freely available programming language, intended primarily for efficient execution can be used to record succinct specifications, some of which may be non-executable in principle. The transparent semantics of functional programs make formal reasoning tractable, and as we have seen, it is possible to record proofs and derivations formally, within the language.

By adopting a pragmatic approach to pixel-level image processing we are able to use the functional language for final implementation of image analysis systems, not only for prototyping or "animation" of the specification.

Our approach addresses the three difficulties identified in section 1.3 since, 1) implementations of lazy functional programming languages are readily available (the Miranda system is a modestly priced commercial product, and several interpreters/compilers for Haskell are available free of charge), 2) the overhead (effort and risk of error) of translating from the specification language to the implementation language is eliminated by adopting a common language for both, and 3) specification and implementation can exist as a single integrated document making it easier to maintain consistency through subsequent modifications.

In adopting a single language for specification and implementation there must inevitably be compromises — a Haskell script will never be as accessible as English prose, as succinct as a Z specification or as efficiently executable as a C program. Hopefully we have shown that the compromises need not to too great.

Appendix

A   A brief introduction to functional programming

Recommended text books for those new to functional programming are [3] and [4].

Haskell is a language being defined and implemented by a world-wide group of academics. Particularly active groups are in Glasgow (Scotland), Yale (USA) Chalmers (Sweden).

The Haskell report [7] introduces the language as "... a general purpose, purely functional programming language incorporating many recent innovations in programming language research, including higher-order functions, non-strict semantics [ie lazy evaluation, see below] static polymorphic typing, user-defined algebraic datatypes, pattern-matching, list comprehensions, a module system, and a rich set of primitive datatypes including lists, arrays, arbitrary and fixed precision integers, and floating-point numbers".

Haskell is in fact very similar to Miranda, its key innovation being a consistent approach to overloaded functions and operators, which need not concern us here.

A functional programming language has just one basic operation: the application of a function to arguments; a functional program is fundamentally just one function, from the program's input to its output. Of course the top level function is usually defined in terms of other functions or values. Each function definition looks very much like a mathematical definition, as in the following simple example.

```
> pythag x y = sqrt ( x * x   +   y * y )
> a = 3
> b = a + 1
> c = pythag a b
```

A.1   Referential transparency

A functional program includes no assignments, and so admits no side effects whatso-ever. An expression may be freely replaced by its value, and vice versa, without changing the meaning of the program. [6]. In this way, a functional program mirrors a system of mathematical equations[15], representing a *static* world — the symbol x say, stands for the same value throughout its scope, and a function f applied to x (written $f\ x$) will always yield the same value. This is the principle of *referential transparency* (see eg [6]) which gives mathematics its deductive power — power which is therefore inherited by functional languages. The assignment operator (ie, '=' in C ':=' in Pascal) has no place in mathematics and likewise does not occur in a functional language.

To contrast this situation with conventional *imperative* languages such as C or Pascal, consider the proposition

$$f(x) + g(x) = g(x) + f(x).$$

Taken as a mathematical statement, we have no difficulty in accepting its truth, regardless of the definition of f or g. However, if the above is considered as a Boolean expression in a C program (replacing '=' with '==') then due to the possibility of side effects (intended or accidental) we could not be so confident. As an expression in a functional program however, our mathematical intuition would hold good.

A.2 Extensional and Intensional properties

The *extensional* properties of a function relate to its mathematical mapping only. *Intensional* properties relate to the algorithm by which the function is computed. For example, in Haskell notation

```
> sqdiff x y = x*x - y*y

> sqdiff' x y = (x+y) * (x-y)
```

we can easily show that `sqdiff` is equal to `sqdiff'` because they have identical extensional properties (they compute the same function) but they have slightly different intensional properties.

A.3 Formal transformation

A functional program, just like any conventional procedural program, may be written in many ways which have vastly differing intensional properties and so each form will execute with differing time and space requirements. An important advantage of a functional implementation is that the definitions may be *transformed* using conventional mathematical reasoning into a form which is more efficient on a given architecture. These program transformations (or *reifications*) can be carried out with formal rigor. It may take skill and intuition to recognise the necessary steps, but each step can be *proved* correct by conventional mathematical reasoning. Semi-automated theorem provers are becoming available (see eg [14]) to perform house-keeping tasks and to validate the elementary transformation steps. It is thus possible to move around a space of different algorithmic implementations in a formal way which guarantees that the computed function remains unchanged. In conventional programming the analogous process is known as "optimizing the code" and is a well known cause of programming errors.

A.4 Lazy evaluation

Lazy evaluation means that all sub-expressions in a program are not evaluated until and unless their value is needed, and are evaluated at most once. This is a great advantage, allowing programs to be written in terms of notionally infinite, or partially undefined structures. It is often possible to write a function in way which is simple and transparent, but which appears computationally naive; lazy evaluation will, in many cases, result in efficient execution never-the-less. For example, a package to compute the terms of a Taylor expansion for a function can be written to return the infinite list of *all* terms; only those terms actually used by the calling program will be evaluated.

References

[1] J R Abrial. *The Specification Language Z: basic library.* Programming Research Group, Oxford University, 1980.

[2] H Alexander and V Jones. *Software Design and Prototyping Using me too.* Prentice Hall, 1990.

[3] R Bird and P Wadler. *Introduction to Funtional Programming.* Prentice Hall International, 1989.

[4] A J T Davie. *An Introduction to Functional Programming Systems Using Haskell.* Cambridge University Press, 1992.

[5] Peter Henderson. Functional programming, formal specification, and rapid prototyping. *IEEE TRANS on Software Engineering,* 12(2):241–250, 1986.

[6] C. A. R. Hoare and J. C. Shepherdson, editors. *Mathematical Logic and Programming Languages.* Prentice-Hall, 1985.

[7] Paul Hudak et al. *Report on the programming language Haskell (V1.1).* University of Glasgow, 1991.

[8] C B Jones. *Systematic Software Development Using VDM, 2nd ed.* Prentice Hall, 1990.

[9] Mark P Jones. An introduction to gofer, 1992.

[10] Leslie Lamport. LaTeX— *A Document Preparation System.* Addison-Wesley Publishing Company, 1985.

[11] P G Larsen and P B Lassen. An executable subset of meta-iv with loose specification. In *VDM'91: Formal Software Development Methods,* pages 604–618, 1991.

[12] J Piper and D Rutovitz. Data structures for image processing in a C language and a unix environment. *Pattern Recognition Letters,* pages 119–129, 1985.

[13] Ian Poole. A functional programming environment for image analysis. In *11th International Conference on Pattern Recognition,* volume IV, pages 124–127, 1992.

[14] Colin Runciman, Ian Toyn, and Mike Firth. An incremental, exploratory and transformational environment for lazy functional programming. *J. of Functional Programming (submitted),* 1992.

[15] D A Turner. Recursion equations as a programming language. In J Darlington and D A Turner, editors, *Functional programming and its applications.* Cambridge University Press, 1982.

[16] D A Turner. An overview of miranda. *Sigplan Notices,* 21:158–160, 1986.

Developing an environment for computer-based automotive suspension and steering systems

J. Robinson and S. Menani Merad
KBSL. 1 Campus Road
Listerhills Science Park
Bradford, West Yorkshire BD7 1HR. U.K.

Abstract

The use of computer controlled systems on road going vehicles is grow-ing. Leading car manufacturers are developing, and in some cases marketing, systems which influence or take control away from the driver. The safety of normal every day drivers and passengers is increasingly dependent on the reliability of such systems.

The CBASS (Computer Based Automotive Suspension and Steering Sys-tems) project is undertaking a programme of work which sets out to research, and prototype, a development and test system for automotive computer con-trol applications.

This paper provides an overview of the CBASS project.

1   Background to the Work

Pressure to add sophisticated control systems to road cars is growing. The need to distinguish products within a competitive market, together with the use of control systems in high profile motor racing applications. means that such systems are likely to become widespread within a relatively short period of time.

On a more technical level, the use of control systems in vehicle suspension sys-tems can overcome problems experienced with more traditional suspension designs. The ideal suspension design. i.e. properly designed double wishbone suspension sys-tems, is unlikely to be offered on average road cars by mass market manufacturers. Such systems are relatively expensive and load space intrusive. The more common solutions, such as Macpherson Strut front suspension, suffer from various problems but are an inevitable consequence of commercial pressures.

It is therefore likely that some form of "intelligent" suspension medium, to min-imise the performance deficiencies of these engineering compromises, will become common place in future mass market road cars.

Such systems will also find application across a wide range of vehicles, not just the performance car market. Environmental pressures will also have an impact. For example, among the studies into more environment-friendly cars have been projects which aim to reduce drag, and hence fuel consumption, by controlling ride height to cope with load variation.

2   Safety Concerns

In any but the most simple automotive control systems. the reliability of the software component is central to the reliability of the system as a whole. Hardware fail safe devices will stop systems failing completely. e.g. stopping an active ride suspension system collapsing. "bottoming" the car. However. it is unlikely that any hardware fail safe device will stop an adaptive suspension system making an incorrect "decision". e.g inducing understeer when it should be inducing oversteer.

The authors suspect that this fact is not currently widely accepted within the automotive sector. In addition. commercial pressures may mean that some traditional aerospace techniques. such as software redundancy. may not be readily acceptable in automotive sectors.

The safety and reliability of these applications are critical issues for the automotive industry. This is particularly true in systems such as [Zom92] in which the control system is intended to increase the overall reliability of the vehicle. Some systems now coming to market will, for example. apply the vehicle's brakes when the control system decides that the vehicle is about to enter an unsafe state. Saab for example [Sun92] are selling cars equipped with a system which stops wheel spin during cornering by automatically shutting down the throttle. and in the case of manual gearbox models, applying the brakes during cornering.

To further complicate matters. there are significant advantages in ensuring that suspension design continues to be the domain of vehicle dynamicists. The use of domain specialists in the production of control systems means that safety-related software development techniques need to be readily accessible to non software specialists. The commonly proposed safety critical related techniques. such as formal methods, are undoubtedly required in this domain. However. they must be presented as part of an environment which is acceptable to the target user. The CBASS environment will therefore have "usability" as a prime requirement.

3   The Proposed Development Environment

Figure 1 provides a diagrammatic overview of the proposed development environment. This environment will aim to provide support for safety critical related technologies within a framework which is acceptable to the end user. specifically a vehicle dynamicist. This framework will combine advanced software development techniques and sophisticated vehicle test rig technology to provide users with an integrated. and hopefully seamless. development and test environment.

Full development of the proposed environment to "production quality" is of course a major undertaking. The overall aim of the CBASS project therefore is to research the requirements for the proposed system. and to prototype key elements of the environment in order to demonstrate its feasibility. In order to demonstrate the true potential of the environment. the deliverables from the project will include actual prototype vehicle systems. It is hoped that these prototype systems will be developed to the status of drivable systems which can be used as demonstrators for the CBASS technologies.

The proposed CBASS environment aims to find errors in a vehicle control application as early as possible in its development lifecycle. Central to the CBASS concept is the aim to provide a development environment which is perceived as appropriate by an engineer, for example a vehicle dynamicist, developing a control application. It cannot be assumed that such an engineer would find the existing proposals from the software community, such as formal methods, acceptable. The CBASS environment will therefore attempt to provide support for such techniques, but in a manner which will be acceptable to the target users. This may involve hiding the presence of these techniques from the user.

One of the major goals of the proposed environment is to reduce the current dependence on intensive live testing of vehicle systems. Manufacturers are heavily reliant on the use of test tracks and race tracks to develop new products. The intention in the CBASS project is to provide "front-end" techniques, in the form of computer-based testing and simulation, and static test rig facilities, to improve the existing development lifecycle. It should be noted that the CBASS project is not suggesting removal of the on-vehicle testing process. Indeed, on-vehicle testing of applications forms an important part of the CBASS project itself since the consortium members wish to demonstrate the CBASS environment across a complete product lifecycle. Rather, the suggestion is that, by improving the front end design process with simulation and similar capabilities, the test load can be reduced and can as a result become more focussed.

Built into the environment, therefore, is the concept of validation and verification throughout the development cycle. This will include "animation" or "simulation" of the application specification on the development workstation. Simulation capabilities will provide a feedback loop at this stage of an application development, allowing a design engineer to refine a design prior to committing this to metal.

A static test rig is also proposed as part of the CBASS environment. Such a rig would in effect be the control application equivalent of a wind tunnel, providing the facility for extended tests of applications under controlled conditions. The project will need to establish the nature and structure of a rig capable of providing such capabilities. The rig will provide a further level of feedback prior to committing a system to a vehicle. This feedback may lead to a change in the specification of the system in which case the effects of any changes may be simulated prior to re-testing on the rig. The CBASS simulation capabilities should therefore be tightly integrated with the test rig, allowing lessons learnt on the rig to be fed back into the development system in order to improve the accuracy of further simulations.

The CBASS environment therefore aims to provide an integrated environment which offers several mechanisms which will reduce the effort required to undertake effective verification and validation of automotive control applications.

4   The Proposed Software Technologies

4.1   Object Orientation - A Natural Solution ?

In order to provide some kind of structure to what could be very complex control systems, there exists a need for a suitable system decomposition technique. Object oriented techniques provide one possible approach to providing the main structuring element of the CBASS environment. Object Oriented techniques offer support for a highly modular structure which supports a building block approach to control system construction.

In addition, an Object Oriented approach to automotive control systems seems very natural. Control systems tend to contain easily identifiable real world "objects" such as sensors, actuators, valves etc. For example, a brief examination of the Active Rear axle Kinematics on the BMW 850i [Car91], immediately produces the following list of physical resources, each of which could be considered a "candidate" object:

- wheel speed sensor

- hydraulic pump

- warning light

- steering wheel angle sensor

- speedometer

- actuator

- pressure supply

Within the object oriented software development paradigm, each of these real world entities would naturally be associated with an object, or object class, within the control software.

Unfortunately, there is a diversity of opinions on what constitutes an Object Oriented approach. The decision to use an Object Oriented approach for CBASS is therefore far from straightforward. There must also be a decision as to which kind of Object Orientation is appropriate for the application in question.

4.2   Which kind of O.O.?

A software engineer with Ada language experience will have a significantly different view of what constitutes an Object Oriented system from a software engineer more used to Smalltalk. Indeed, attempts to compare these two approaches, sometimes referred to as Object Oriented Design (OOD) and Object Oriented Programming (OOP), have resulted in entirely different conclusions. Rosen [Ros92] for example concludes that an Ada style approach is Object Oriented whilst Cook [Coo86] concludes that Ada supports none of the requirements for an Object Oriented language.

Consider for the moment just one of the issues related to Object Orientation, as highlighted by Hall [Hal88]:

"the specialisation of components through parameter substitution, added material or perhaps even modification to existing material".

There exists two principal approaches to achieving such specialisation, the inheritance mechanisms typically found in OOP languages and methods, and the concept of genericity found in OOD languages and methods. Myer [Mye88] provides a useful comparison of the two approaches. Support for each approach amongst the different Object Oriented camps is very strong, almost to the point of being religious.

A typical Smalltalk engineer would consider the concept of inheritance as central to OOP. If inheritance is not supported then a language, method or environment is not object oriented. Yet the much more straightforward generic facility of Ada is not seen as a mandatory feature of OOD. Despite being so much simpler than most inheritance mechanisms, it is often banned by Ada Codes of Practice for safety critical applications. It would appear that the provision of an inheritance mechanism in the next version of Ada, Ada 9x [Tlu91], is doing very little to bring about a convergence of opinion.

As a result, there is as yet no clear industry position on the use of genericity or inheritance for the development of systems which need to meet stringent reliability and safety requirements.

This issue is key to the whole structure and appearance of the proposed environment. It is central to the choice of object retrieval and interconnection mechanism for example. Since these mechanisms are inextricably linked to the use of adaptive techniques (section 2.2) and formal representations (section 2.3), it can be seen that the choice of Object Oriented approach will effect the entire appearance and concept of the CBASS environment.

4.3   What kind of O.O. Lifecycle

Much of the work undertaken on Object Orientation and re-use assumes some form of lifecycle consisting of requirements analysis, design etc. Indeed some work, such as [Bur92], has been oriented towards choosing an effective lifecycle. What is not clear to the authors at this stage is what kind of lifecycle, if any, a vehicle dynamicist works within. Clearly, establishing the lifecycle which needs to be supported will be an important aspect of the early stages of the research.

5   Adaptive Techniques

Many control systems need to operate in environments which are subject to subtle changes. A control system will often need to be sensitive to such changes, displaying behaviour which adapts to circumstance. Adaptive techniques, such as qualitative reasoning, deal with vague or qualitative situations, particularly in systems where there are complex interactions between system components.

5.1   Why use Adaptive Techniques?

One of the major problems in automotive control is the ability to deal with qualitative or vague properties. This causes problems when trying to model the system [Lei88] since:

- According to circumstance, the same system could manifest itself in different ways, leading to a poor understanding of the process to be controlled

- The element of uncertainty could result in having to estimate the values of some parameters in the model

There is a danger that, in applying traditional control techniques to a qualitative situation the soundness of the underlying technique is lost. The pressure to maintain high levels of reliability therefore leads to a need for alternative approaches more suited to qualitative situations.

In control systems knowledge is uncertain and incomplete. Hence, the need for techniques which are close to human reasoning providing representation and reasoning mechanisms to explicitly support such knowledge. Adaptive techniques provide the right level of abstraction to deal with the qualitative and vague properties of control systems. Fuzzy logic could be regarded as an appropriate technology to represent the qualitative properties of the system.

Consider the use of fuzzy logic in an automatic braking system described by Aurrand-Lions et al in [Aur91]. This system, which has a prototype implementation installed on a Citroen XM test car, is designed to accept data and instructions from road side beacons. A beacon located at a stop sign or a red traffic light will instruct the vehicle to stop. If the driver does not respond, the braking system automatically brings the vehicle to a halt.

Fuzzy control theory is based on fuzzy logic which enables deduction using rules such as the following:

Condition 1: If the speed is very high and
Condition 2: if the deceleration is slightly positive
Action: then braking pressure is moderately increased

The use of fuzzy concepts simplifies the implementation of this sort of system. The terms "very high", "slightly positive" and "moderately increased" are fuzzy qualifiers which are modelled by fuzzy sets [Zad65, Zad73]. A fuzzy set is a set which members have grades of membership ranging between 0 and 1. Hence, each fuzzy set is formally associated with a membership function. For instance, speed could be associated with many qualifiers (very low, low, normal, high, very high) each of which is modelled by a fuzzy set. Since, there is a mathematical representation of those fuzzy terms, one can combine propositions using logical connectives and thus infer conclusions from fuzzy rules as those above.

This approach is very flexible, particularly when complex combinations of rules are applicable to a single input or output. For example, the action being controlled in the braking example could also be influenced by other non-crisp inputs, such as the prevailing weather conditions and the condition of the road surface.

As a result of the advantages provided by fuzzy techniques, significant interest in these techniques has been generated in the automotive industry. Other examples of the use of fuzzy logic in an automotive domain can be found in Ikeda et al [Ike91], Takahashi [Tak91] and Terano et al [Ter91]. Although the projects report positive results from the use of these techniques, no mention is made of how safety requirements had been satisfied, or even analysed.

It has even been proposed in [Lee91] that fault tolerance could be implemented by using a fuzzy controller to switch out faulty channels in a dual redundant control system. No mention is made of how the reliability of the fuzzy controller itself could be assured however.

A range of technologies show some potential for use in this domain. Neural Networks for example introduce the possibility of training a system rather than programming it. Rather than defining the behaviour of a system, a neural network will deduce decision strategies directly from accumulated data [Hud91]. "Training" an automotive control system would involve providing it with historical data from a test rig or simulation environment, perhaps similar to the proposed CBASS test rig.

Although this concept may appear initially to be somewhat outlandish, an example of this kind of approach can be found in the automatic braking example in [Aur91]. As part of the activity to define the fuzzy sets needed in the system, the project team used a simulator to perfect the performance of the system.

5.2   Choosing an appropriate level for adaptive behaviour

There exist several possible approaches to building adaptive behaviour into an object oriented environment:

- adaption at object level
- adaption at object interconnection level
  - using a fuzzy controller
  - using message modifiers

5.2.1   Adaption at Object Level

One can envisage restricting the CBASS environment such that adaptive techniques may only be used as part of the implementation of an object. This has the advantage of "hiding" adaptive behaviour behind a well-defined interface. Allowing an object to have a fuzzy implementation whilst exporting a crisp interface.

Figure 2 shows the automatic braking system in the context of an object oriented model. Note that the presence of fuzzy logic in the system is only obvious when the implementation detail of "Brake" is revealed. At the object interconnection level no sign of fuzziness can be found.

5.2.2   Adaption at Object Interconnection Level

It may be more practical or appropriate to raise fuzzy and adaptive concepts to the level of object interconnections.

Figure 3 shows a modified version of the automatic braking system, in which a Fuzzy Controller "object" has been introduced to encapsulate the fuzzy decision making. This is a very common approach in fuzzy control systems, where a central fuzzy controller will interact with several interface objects.

It could be argued that the separation of the fuzzy controller from the brake object represents a useful separation of concerns, i.e. the brake object deals with the physical details of controlling the actuator whilst the "thought process" which "decides" how hard to brake is separated and encapsulated elsewhere. However, this solution breaks a fundamental rule of Object Orientation since the fuzzy controller has a distinctly functional feel to it.

One possible solution to this is to think of the fuzziness applying not to objects in the system but to message paths in the system. In other words, a message sent from one object to another will proceed along a path, during which it is modified according to some fuzzy rules. It could also be argued that this is no different to the concept of placing constraints on relations between object classes, as is common practice in some Object Oriented Analysis techniques. A possible notation for this approach, using a Message Modifier, is shown in Figure 4.

This is an unusual and potentially quite flexible approach. Its impact on the predictability of system behaviour however, together with possible implementation techniques, presents a fascinating area of research.

In raising the fuzziness to the level of object interconnections, it could be argued that we could become more aware of the fuzziness in the system, allowing us to assess its impact on safety. For example, the constraints:

Do not apply brakes if we are cornering quite quickly

or

Don't apply brakes if the vehicle is sliding

if hidden inside the implementation detail of a brake object would not be obvious when viewing the system as a whole. However, if stated explicitly as part of object interconnections they become obvious at a system level.

Deciding whether or not the implementation of these constraints as Message Modifiers is an effective and appropriate way of building the system will be an important aspect of the CBASS project.

5.2.3 Choosing an approach

The choice between these two approaches to fuzziness, or indeed the third possible approach of a combination of the two, is driven by a number of factors:

- The requirements of the application domain, e.g. is adaptive behaviour needed at the level of object interconnections.

- The problems of hiding adaptive behaviour behind a well defined interface, i.e. it may not be possible to stop adaptive objects interfering with object interconnections.

- The need to maintain an object oriented viewpoint and to avoid the temptation of producing a "functional object" such as a fuzzy controller.

- The need for CBASS to reflect the implementation domain. For example, it is possible for fuzzy controllers to be implemented in hardware, giving significant performance gains over software implementations. In this case fuzziness at the object interconnection level may be inevitable. Indeed, it could be argued that the fuzzy controller is, in this case, a valid real world object.

- Constraints on the implementation technologies available. It should be noted that a fuzzy Object Oriented programming language has been proposed by Yasunobu et al [Yas91, Yam91, Ino91].

Whatever the approach taken, adaptive techniques must be able to operate within defined safety constraints.

The CBASS project will seek to identify the techniques and approaches to adaptability which are most appropriate for the development of safety critical systems. The project will pay particular attention to the need for imposing safety related constraints on such adaption.

6 Applying Formalism to the CBASS Environment

In any safety critical domain there will be pressure to show that the developers have used techniques to ensure appropriate levels of reliability and confidence. Formal methods have potential to provide the basis for a safety oriented development environment. However, any such method must be offered in a style and notation which is acceptable to the end user. There is of course a potential philosophical clash between the proposed adaptive techniques and the use of rigorous formal notations which must also be resolved.

A number of issues need to be addressed by the CBASS project:

- How should the development process be partitioned to cater for the qualitative and safety properties of the system.

- What form should the formal representation take. e.g. should it be in an existing notation such as Z or should it be in a new, possibly domain specific notation?

- Should the formal representation be simply some form of intermediate code, hidden from the target user, or should it be visible to, and possibly manipulatable by, the target user?

- Should the formal representation be used simply to define safety constraints, or to define system functionality as well? It may for example be possible to define system functionality within the terms of Object Oriented and adaptive techniques, whilst imposing system wide safety constraints using a formal notation.

- Should the formal notation be used as part of an object's interface specification? Should it be used as part of object interconnections? Should it be used to prove the correct implementation of an object's interface?

- Is a formal specification of the adaptive behaviour of the system feasible?

6.1 Why Use Formalism

A formal specification of a system provides a rigorous and unambiguous specification of system behaviour [Woo88]. However, formal notations, and some so called formal methods, only provide a mathematical language. One approach to overcome this problem is the integration of the Object Oriented model with a formal notation. A number of different approaches to this problem have been taken, for example [Hal90, Why90b, Why90a, Duk91, Lan90, Sch87, Mei90, Mer92a, Mer92c, Mer92b]. What is clear from these different approaches is that an integration between a formal notation and an Object Oriented method or language is not only feasible but also brings a number of benefits.

The FOOSA project [Mer92a, Mer92c, Mer92b], for example, has illustrated the concept by formalising an Object Oriented Analysis (OOA) method with the addition of the Z notation. This approach combines the advantages of the formalism with the advantages of the OOA diagrams, without weakening the rigour of the formal notation. It effectively provides a diagrammatic representation of the formal specification.

Halang and Kramer [Hal92] have produced a development environment which combines an application specific graphical language with formal specification and an automated support environment. This system, targeted at process control engineers using programmable logic controllers (PLCs), therefore provides an environment in which the target user works within terms acceptable to themselves, whilst also being supported by formal specifications enabling formal verification and prototyping of specifications.

Halang and Kramer reported an improvement in productivity as a result of the graphical design environment, but noted that formal specification had to be added in order to meet the expected degree of reliability in the resultant control software.

6.2   What kind of formal notation?

Although formalism seems to fit into the CBASS environment in a natural manner, it will be important to consider very carefully what kind of formal notation is needed.

Possible alternatives to the traditional textual style of notation include the use of visual programming techniques [Dra92] to make a formalism more acceptable, as in Roberts and Samwell [Rob89].

Dick and Loubersac [Dic91] has also proposed the use of a graphical notation to make formalism more acceptable.

7   Formal vs Fuzzy

It may seem contradictory to attempt to build any system which includes both formal and fuzzy concepts. Indeed the two are often presented as if they are worlds apart.

It is important to realise that, in spite of their name, fuzzy theories are mathematically based. Indeed many of the theories are now mature and well established, such as fuzzy sets [Zad65], fuzzy logic [Zad78] and fuzzy control theory [Ped89]. Work has also been undertaken on the mathematical specification of fuzzy logic programming languages [Mer91].

Using fuzzy concepts in a safety critical domain will still present some problems however. For example, the membership functions which represent the fuzzy sets are subjective in nature. In a safety critical domain these would need to be evaluated precisely and made as objective as possible. Potential techniques for doing this are already available [Sme88].

7.1   Choosing an Appropriate Formal Notation

Another important aspect to the research will be the choice of formal notation. The ideal solution would use a single notation across the environment.

However, it may be impossible to identify one single notation which is sufficient, given the variety of technologies being proposed for the CBASS environment. It could be that the use of more than one notation is more appropriate, although it would be preferable to keep the number used to a minimum. It would also be necessary to establish formal rules to link the formal notations used.

Alternatively, if using a number of notations turns out to be unpractical, the creation of a new notation that would cover all the aspects of the CBASS system is a possibility. Wherever possible however an existing, preferably well accepted notation would be used.

The initial proposal for CBASS is to apply a technique similar to that used in FOOSA to the object oriented model supported by CBASS. In other words, take the object oriented representation of an automotive control system and formalise it by using it to generate a formal representation.

However, it is anticipated that user feedback could change this approach significantly. The project team are keen to see what kind of formalism appeals to the potential target users of the CBASS environment.

8 Applicability of the CBASS Environment

Although focussed on automotive applications, it is hoped that the project will deliver results which will have applicability to other safety critical applications such as process control and medical applications.

Indeed, it is possible that the CBASS environment could provide the base for a generic environment, tunable to specific problem domains.

To this end, the project will investigate whether CBASS will conform to generic standards such as those emerging from IEC WG10 as well as automotive oriented standards such as those emerging from the DRIVE programme.

9 The Consortium

Lead partner of the CBASS consortium is KBSL, a specialist software engineering company who have been involved in safety critical related technologies since formation in 1985. Further software expertise is being provided by Design in Computing, a software house specialising in industrial and control applications.

The automotive sector is represented by two partners. R&SS, a specialist manufacturing and development company involved principally in motor racing applications, and Engineering Research & Application Ltd (ERA). ERA are a leading centre for the development and test of vehicles and automotive components.

The CBASS consortium therefore includes representatives of the defence, automotive, aerospace and industrial control sectors. Early discussions regarding the CBASS project have highlighted several areas of diversity of software technology between these different sectors. The consortium partners expect to gain a great deal from exposure to problems and solutions experienced in industries other than their own.

10 Further Cooperation Sought

The most important basic aim of the CBASS project is to ensure that the results of the project are pragmatic and targeted real user needs. It is therefore important for the consortium members to establish and maintain effective communication links with those in the automotive sector involved in the development. use and assessment of automotive control systems.

The project will include an extensive market research study aimed at monitoring current development methods and establishing the requirements for future development environments. Any interested parties are encouraged to contact the authors to establish an initial contact.

11 Acknowledgements

The authors would like to thank Chris Groves. of G.I.L. Design. for his contribution on vehicle dynamics. as well as for many fascinating and enlightening discussions.

Vehicle
Specification
(or request for
change)

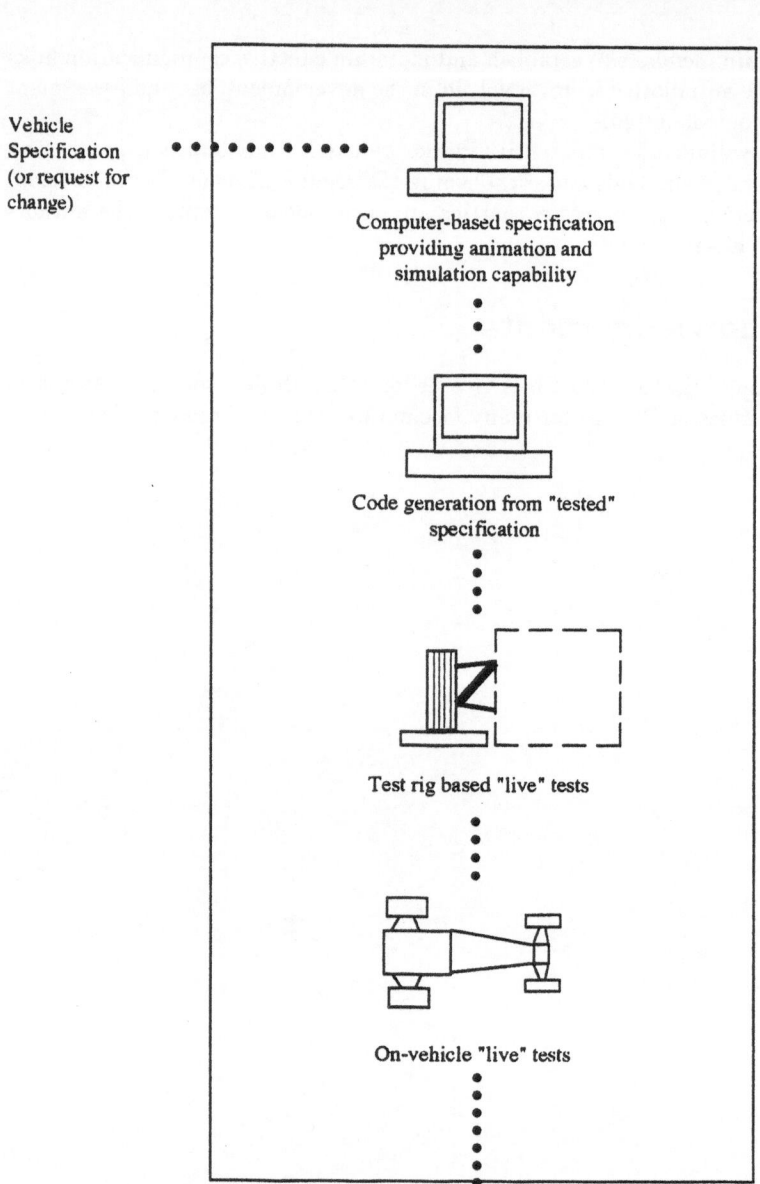

Computer-based specification
providing animation and
simulation capability

Code generation from "tested"
specification

Test rig based "live" tests

On-vehicle "live" tests

Delivered Product

Figure 1
CBASS Overview

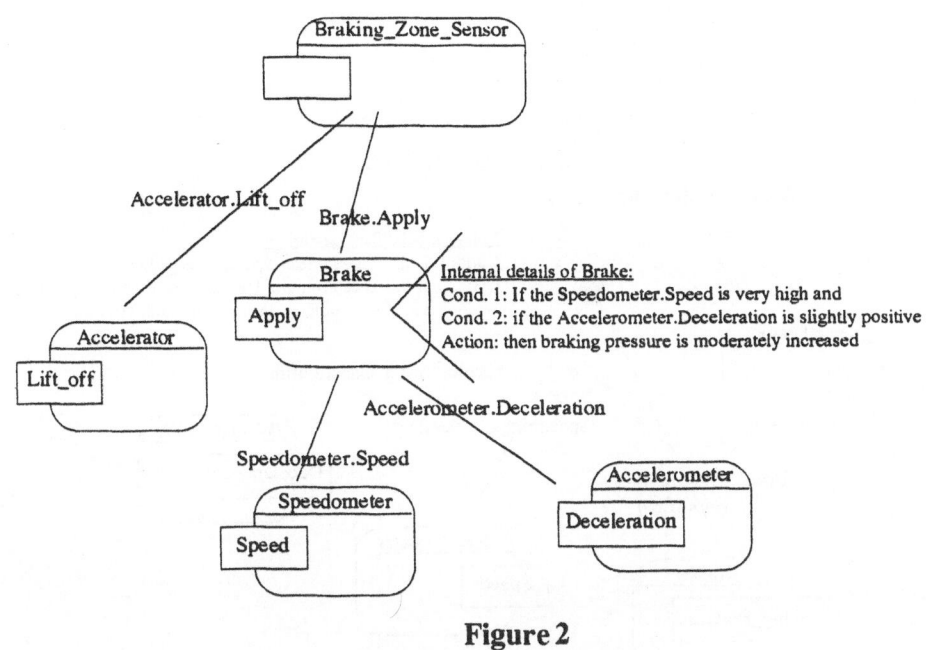

Figure 2
Object Level Fuzziness

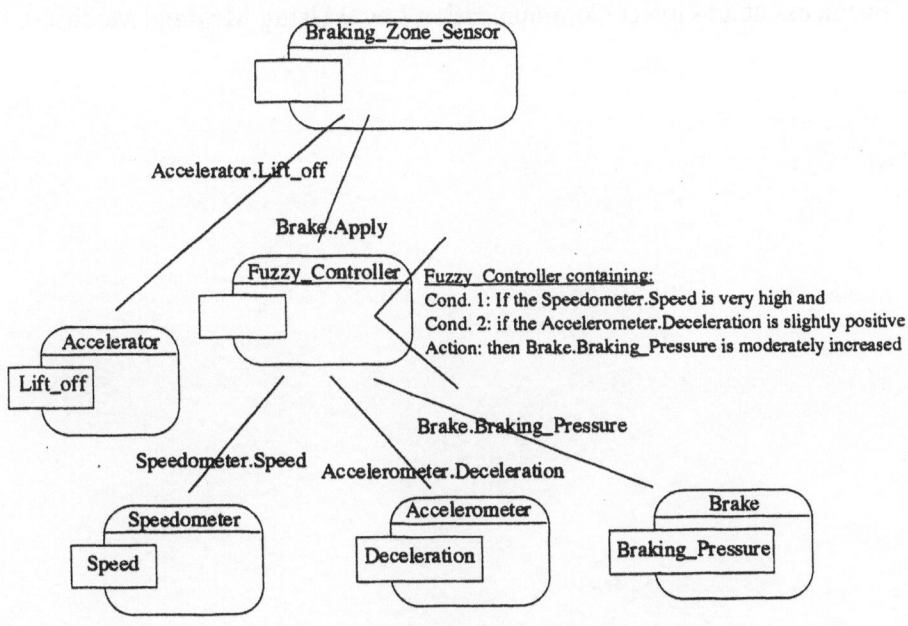

Figure 3
Fuzziness at an Object Communication Level Using a Fuzzy Controller

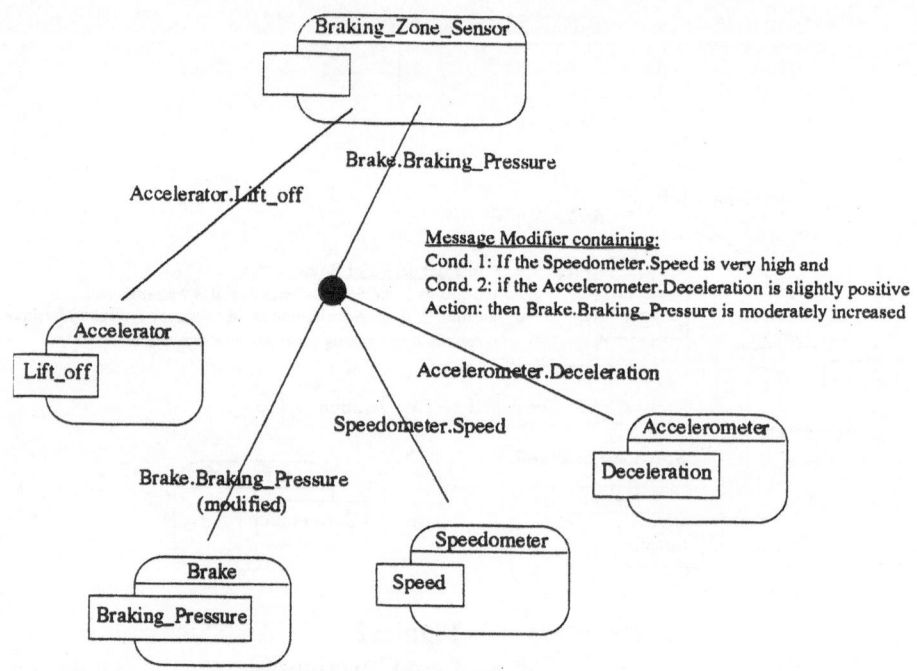

Figure 4
Fuzziness at an Object Communication Level Using Message Modifiers

References

[Aur91]   J P Aurrand-Lions, L Fournier, P Jarri, M de Saint Blancard, and E Sanchez. Application of fuzzy control for ISIS vehicule (sic) braking. In *IFSA '91, Brussels,* 1991.

[Bur92]   E Burd. The spiral model and object-orientation: a path towards successful reuse. In *KBSL conf. on Requirements and Design Methods for Object Oriented Environments,* June 1992.

[Car91]   BMW Follow the High Technology Route. *Car Design & Technology,* (2), August/September, 1991.

[Coo86]   S. Cook. Languages and object-oriented programming. *Software Engineering Journal,* vol. 1, No. 2, pp.73–80, 1986.

[Duk91]   R. Duke, P. King, G.Rose, and G. Smith. The Object-Z specification language. In T. Korson, V. Vaishnave, and B. Meyer, editors, *Technology of Object Oriented Languages and Systems: TOOLS 5.* Prentice-Hall, 1991.

[Dic91]   J. Dick and J. Loubersac. A visual approach to VDM: Entity structure diagrams. Technical Report DE/DRPA/91001, Bull Corporate Research Centre, Louvensiennes, France, January 1991.

[Dra92]   N. Drakos. A case for object orientation and visual programming. In *KBSL conf. on Requirements and Design Methods for Object Oriented Environments,* June 1992.

[Hal88]   P. A. V. Hall. Software components and re-use. *Software Engineering Journal,* vol. 3, No. 5, pp.171, 1988.

[Hal90]   A. Hall. Using Z as a specification calculus for object oriented systems. In D. Bjorner, C.A.R. Hoare, and H. Langmaack, editors, *VDM and Z - Formal Methods in Software Development, Lect. Notes in Computer Science,* VDM-Europe, December 1990. Springer Verlag.

[Hud91]   D.L. Hudson, M.E. Cohen, and M.F. Anderson. Use of neural network techniques in a medical expert system. *Int. J. Intell. Syst,* 1991.

[Hal92]   Wolfgang A Halang and Bernd Kramer. Achieving high integrity of process control software by graphical design and formal verification. *Software Engineering Journal,* vol. 7, No. 1, pp.53–64, January 1992.

[Ike91]   H. Ikeda, Y. Hiramoto, N. Kisu, and H. Kahashi. A fuzzy processor for a sophisticated automatic transmission control. In *IFSA '91, Brussels,* 1991.

[Ino91]   Y. Inoue, S. Yamamoto, and S. Yasunobu. Fuzzy set object: fuzzy set as first-class object. In *IFSA '91, Brussels,* 1991.

[Lan90]   K. Lano. An object-oriented extension to Z. In *Proc. of the Z User Meeting,* 1990.

[Lee91]   H. Lee, A. Bien, and Y. J. Cho. Time weighted fault-tolerant control using fuzzy logic. In *IFSA '91, Brussels*, 1991.

[Lei88]   R. Leitch. Software components and re-use. *Software Engineering Journal*, vol. 3, No. 5, pp.171, 1988.

[Mei90]   S.L. Meira and A.L.C. Cavalcanti. Modular object oriented Z specification. In *Proc. of the Z User Meeting*, 1990.

[Mer91]   S. Menani Merad. *A Theory for a Fuzzy Logic Programming System*. PhD Thesis, University of Bradford, 1991.

[Mer92a]   S. Menani Merad. FOOSA (Formal Object Oriented Analysis), phase I. Technical Report KBSL technical report under contract RSRE 1C/6129. DRA Electronics Division, RSRE, Malvern, UK, May 1992.

[Mer92b]   S. Menani Merad. FOOSA (Formal Object Oriented Analysis), phase II, report 2. Technical Report KBSL technical report under contract RSRE 1C/6129, DRA Electronics Division, RSRE, Malvern, UK, May 1992.

[Mer92c]   S. Menani Merad and J. Robinson. FOOSA (Formal Object Oriented Analysis), phase II, report 1. Technical Report KBSL technical report under contract RSRE 1C/6129. DRA Electronics Division, RSRE, Malvern, UK, May 1992.

[Mye88]   B. Myer. *Object-oriented Software Construction*. Prentice Hall Intl. 1988.

[Ped89]   W. Pedrycz. *Fuzzy Control and Fuzzy Systems*. John Wiley, 1989.

[Ros92]   J. P. Rosen. *Object-oriented Paradigms: OOD versus Inheritance*. Ada Yearbook 1992. Chapman & Hall, 1992.

[Rob89]   M. Roberts and P. M. Samwell. A visual programming system for the development of parallel software. In *Proc of the 2nd Intl Conf on Software Engineering for real-time systems, IEE conf. publication No 309*. 1989.

[Sme88]   P. Smets. The measure of the degree of truth and of the grand of membership. *Fuzzy Sets and Systems*, vol. 25, pp.67–72, 1988.

[Sch87]   S.A. Schuman and D.H. Pitt. Object oriented subsystem specification. In L. Meertens, editor, *Program Specification and Transformation*. Elsevier Science, North Holland, 1987.

[Sun92]   Saab beats wheelspin with a high-tech sensor. *The Sunday Times*, February 1992.

[Tak91]   H. Takahashi. A method of predicting the driving environment from the driver's operational inputs. In *IFSA '91, Brussels*, 1991.

[Ter91]   T. Terano, S. Masui, and K. Nagava. Fuzzy control of bulldozer (sic).
           In *IFSA '91, Brussels*, 1991.

[Tlu91]   A. Tlusty-Sheen. Object orientation and Ada 9x - influences and con-
           sequences. In *MSET Conference*, October 1991.

[Woo88]   J.C.P. Woodcoc and M. Loomes. *Software Engineering Mathematics*.
           Pitman, London, 1988.

[Why90a]   P.J. Whysall and J.A. McDermid. An approach to object oriented spec-
           ification using Z. In J. E. Nicholls, editor, *Z User Workshop*, 1990.

[Why90b]   P.J. Whysall and J.A. McDermid. Object oriented specification and
           refinement. In *4th Refinement Workshop*. Springer Verlag, 1990.

[Yas91]   S. Yasunobu. A proposal for the architecture of a fuzzy computer. In
           IFSA '91, Brussels, 1991.

[Yam91]   S. Yamamoto, Y. Inoue, and S. Yasunobu. Object-oriented fuzzy set
           manipulation - internal data object-oriented fuzzy set manipulation -
           internal data. In *IFSA '91, Brussels*, 1991.

[Zad65]   L.A. Zadeh. Fuzzy sets. *Information & Control*, vol. 8, pp.338–353,
           1965.

[Zad73]   L.A. Zadeh. Outline of a new approach to the analysis of complex
           systems and decision processes. *IEEE Transactions on Systems, Man
           and Cybernetics*, vol. 3, No. 1, pp.28–44, 1973.

[Zad78]   L.A. Zadeh. PRUF - A meaning representation language for natural lan-
           guages. *International Journal of Man-Machine Studies*, vol. 10, pp.395–
           460, 1978.

[Zom92]   A. Zomotor, H. Leiber, S. Neundorf, K. H. Richter, and K. H. Buechle.
           Mercedes-Benz 4matic, an electronically controlled four-wheel drive sys-
           tem for improved active safety. *Daimler-Benz AG, SAE 861371*, 1992.

The Practical Application Of Formal Methods To High Integrity Systems

The SafeFM Project

Peter Bradley
AEA Technology, SRD
Cheshire, UK

Linda Shackleton
GEC Avionics
Rochester, UK

Victoria Stavridou
Royal Holloway and Bedford New College
Egham, UK

Abstract

The SafeFM project is a collaborative research initiative involving AEA Technology SRD, GEC Avionics and Royal Holloway and Bedford New College. The project aims to provide guidance on the cost effective requirements capture, development and assessment of high integrity systems.

These guidelines, which will be based upon practical experience, aim to build upon currently available technology by selectively applying formal development techniques. In particular the project aims to identify how disparate requirements can be defined, these requirements effectively developed and the requirements capture and development assessed. SafeFM aims to effectively bring together technology which is typified within emerging national and international standards.

1. Introduction

The software development process can vary from the fully formal proof oriented style (e.g. the ESPRIT BRA Procos Project [1] where formally expressed requirements are refined through provably correct transformations) to the informal, but rigorous production of code which satisfies its requirements. At present the formal approach is not well established whilst the semi-formal approach can make the dependability evaluation of large complex systems onerous.

The complete formal development approach receives considerable attention within the academic world. Indeed, this formal mathematical approach to specification and the subsequent development of software systems may enable larger and more complex systems to be developed in the future. However, although a more formal approach is seen to be necessary and standards are mandating their use [2], there is still considerably less use being made of formal methods in industry. We believe that one of the major reasons for the lack of a widespread use of formal methods can be attributed to the lack of a well defined migration path showing how presently available formal methods and tools can be used effectively to enhance current semi-formal development approaches.

This project aims to provide guidance on the cost effective requirements capture, development and dependability assessment of high integrity systems. These guidelines, which will be based upon practical experience, aim to build upon currently available development and assessment technology by selectively applying formal development techniques. In particular the project aims to identify how disparate requirements can be defined, these requirements effectively developed and the dependability of these requirements and development assessed.

The essence of our development process is that it will be a realistic and practical enhancement to current best practice. This means that it will be constrained by a number of factors such as current standards and practices, tool support, quantification and dependability assessment.

This consortium brings together the extensive experience of AEA Technology SRD[1] in the use of software dependability analysis and dependent failures, the industrial experience of GEC Avionics in the development of high integrity and safety critical avionics systems and the experience of the Royal Holloway and Bedford New College in the state of the art formal methods research.

2   Research Areas

In this section we discuss the current issues related to the procurement of dependable software based systems. We identify three closely related areas:

o   Software requirements analysis and specification

o   Software development (in particular focusing upon the migration path from traditional approaches to formal methods)

o   Software dependability assessment

---

[1]   Project details can be obtained from: Peter Bradley, AEA Technology, SRD, Wigshaw Lane Culcheth. TEL: 0925 254382, FAX: 0925 254539

2.1   Software Requirements Analysis And Specification

Current development methods allow for the analysis and specification of systems that exhibit well-known and understood features. As the breadth of applications of software controlled systems widens and the complexity of these systems increase, there is a demand to extend the expressiveness and increase the formality of these methods. This would enable the characteristics of new systems to be defined accurately and provide a means to validate their performance.

To be able to meet this demand a number of issues need to be considered:-

o   Analysis of all software and system requirements need a number of different viewpoints to be taken. This requires various models of the system such as the design model, the control model and the redundancy management model to be considered.

o   A number of analysis techniques such as fault tree analysis, real-time logic and timed petri-nets are being used in limited contexts. However, system wide techniques that allow consideration of the control system rather than just of the software in isolation require further development.

o   The technical challenge of requirements analysis and specification can be enhanced through a more coherent approach involving the co-operation of systems and software engineers supported by a formal notation which will enable greater consistency and accuracy of system specifications.

2.2   The Migration Path Of Formal Methods Into Existing Best
Practice

Formal methods offer considerable benefits for the production and assessment of high integrity systems. However, the adoption of formal methods has been slow. This can be attributed partly to the following issues:

o   Large investment in staff training is required
The application of formal methods requires discrete mathematics and logic background and many of the current generation of software engineers lack this knowledge.

o   Lack of robust tool support for formal development
Tool support is essential for constructing error free specifications. Writing a formal specification without a good syntax and type checker can be likened to writing a computer program without the support of a compiler.

o   Little evidence of a quantified improvement in the final product

o   Few guidelines on how formal methods can be integrated into current best practice
Although research is advancing the theory of formal methods little progress has been made in the practical use of them within the software development process.

Various research directions are currently being pursued which aim to improve the technology of formal methods. One approach is to integrate a structured (semi-formal) method with a formal development method. Although in theory it is possible to outline how two such methods may be combined, in practice there are difficulties associated with scaling up these techniques for large systems. The second approach is to develop new formal notations with integrated toolsets.

2.3 Assessment of Dependable Systems

Large complex systems can be effectively produced using current technology. However, demonstrating the dependability of such systems has scope for improvement. A number of issues are associated with the dependability assessment of high integrity systems:

o   The integration of assessment techniques within the development lifecycle

o   The cost of dependability evaluation can be comparable to that of the development

o   There are few guidelines and little consensus on the most appropriate method for evaluating the dependability of a complex system (although this consensus is emerging through European wide initiatives)

Typical of the state-of-the-art is the ESPRIT II DARTS [3] project which aims to validate the cost effectiveness of a number of currently available/emerging assessment models. This is being achieved by applying these techniques, within a controlled environment, to diverse safety critical software based systems.

3   Objectives Of SafeFM

The project aims to address the issues related to the requirements capture, development and assessment (as discussed within section 2) by:

o   Adopting a systems-wide approach to identifying requirements and expressing specifications

o   Fortifying existing best software engineering practice with formal methods

o   Extending existing dependability assessment techniques by taking advantage of a more formal development approach

These objectives have been translated into three separate but closely interrelated research streams dealing with requirements capture/specification, development and dependability assessment. Accordingly SafeFM is structured on the following work-streams:

o   Coherent specifications for safety-critical systems
o   Migration paths for formal methods into existing best software engineering practice
o   Dependability assessment

4 Overview Of The Project

4.1 Coherent Specifications For Real Time Systems

This work-stream aims to produce a generic model for the specification of high integrity systems. This will be achieved by research in the following two areas:

o   Identification Of Applicable Models

Although this work-stream is driven by a practical consideration (the production of coherent specifications) achieving this goal requires some fundamental research. We shall examine the issues by examining, in the first instance, the various development stages. We have already identified the need for a systems, a control and design model. We aim to extend the approach with at least one additional model specifying the redundancy executive system [4]. Further models for consideration originate from the areas of hazard and reliability analysis (for instance the SIFT [5] project has used Markov models for the latter).

Clearly the choice of models will be influenced not only by the application domain but also by the assessment principles of the third work-stream. This is because the models used must:

o   reflect assessment concerns such as safety

o   be structured in such a way as to aid the task of assessing the resulting specification

It is believed that the identified models will be generic (as are the control and design models which currently exist) and will be applicable to a wide range of domains.

It is envisaged that the emerging methodological principles will be exercised using case studies of varying complexity. Such case studies will involve theorem proving to show the consistency of the models. We intend to perform such verification experiments involving partly manual and partly machine-aided proof. This aspect will benefit from past and current theorem proving work at RHBNC.

o   Construction Of A Calculus To Represent Models

The second part of the work will construct a simple and elegant calculus for specifying multi-disciplinary system aspects in a unified way. The starting point for this work will be the interval temporal logic with real time durations [6] which has been developed by the ESPRIT BRA ProCos project. We will, however, also investigate the possibility of adapting other notations such as timed CSP, real time logic and Orwellian real time VDM [7] which is currently under development at RHBNC.

It may not be feasible to use the defined calculus for the migration path work-stream of SafeFM, since it will only become available during the later stages of the work. However we expect that the emerging methodology for producing coherent specifications will be independent of any particular notation, and it will, therefore, be possible to use it in the construction of specifications in the migration path work-stream. It is likely that our calculus will be sufficiently different in terms of expressive power and underlying theory from the chosen specification language of the 2nd work-stream to warrant investigating their relationship. Establishing a correspondence between the notations will be useful in both facilitating the use of the methodology and in refining the calculus itself so that it fits better into the migration path framework.

4.2 Migration Path Of Formal Methods

This work-stream aims to show how formal methods can be introduced practically into existing best practice.

o Selective Use Of Formal Methods

Our approach is to look at the use of formal methods at key points in the development lifecycle and assess the benefits to be gained and the cost implications compared to existing techniques. By assessing how we manage issues like complexity and traceability at each phase of the lifecycle with traditional (semi-formal) techniques we can establish what aspects need to be addressed by a formal development approach.

The management of using formal methods is an important issue that has not been addressed by any current work. At present there are no techniques for controlling and structuring large formal specifications. Just as the increasing size and complexity of computer programs resulted in the need for structured methods to help manage and structure the software development process, the increasing size of formal specifications now demand the same attention.

At the specification phase we intend to concentrate on the benefits of a formal specification for safety analysis proofs. Proof is currently a very expensive activity and so careful analysis of which requirements should be proven is necessary. We intend to use software fault tree analysis to help identify the key safety requirements that should be proven.

At the design and coding phase we shall concentrate on the role formal design of specifications can play in the derivation of a good design and code. This will build on our existing work on the use of formal methods with techniques such as Yourdon and HOOD. We intend to determine heuristics for producing specifications at different levels of abstraction and provide guidelines on how they can be used for animation, derivation of test cases and the production of program assertions.

o Integration of Phases

The transition between the phases of the software lifecycle will be investigated to establish how verification and validation can be best achieved.

The transition from the requirement specification into a good design is not a straight forward process, even with the traditional semi-formal approach. However, we intend to identify the features of a formal specification language that will aid this transition. For instance, the language needs to be flexible enough to allow some restructuring of the specification if required, without incurring large penalties in re-working the specification.

The transition from the design to code will also be addressed. The theoretical possibilities include refinement into code [8] or more practically post-hoc verification using a tool such as the SPADE theorem prover.

o   Tool Support

Throughout this work-stream the emphasis will be on the practical use of formal methods. To achieve this, tool support is essential and our approach will be greatly influenced by available tool support. However, we also intend to identify areas where automation would be fairly easy to achieve and would also help to reduce costs. In particular the following areas will be addressed:

o   Automatic generation of structure diagrams of the specification.

o   Automatic generation of SPADE annotations from the formal specification.

o   Automatic code or code template generation from a formal specification.

4.3   Software Assessment

The assessment work-stream will have considerable influence upon the work carried out within work-streams one and two. The purpose of the assessment work-stream is to develop techniques within the dependability lifecycle, reinforcing them with formal methods. This will be achieved as follows:

o   Dependability Within The Coherent Specification

Within this activity we shall investigate the issues associated with defining/assessing dependability requirements within a specification. We will also investigate how a formal specification can be used to fortify the dependability assessment of the specification.

We believe that in order to assess the safety of requirements presented within a complex specification it is necessary to view the specification from a number of different perspectives (eg, fault tree, failure mode/effects and event tree). We intend to review the issues inherent in specifying dependability criteria, identify the important perspectives necessary to specify the dependability requirements for a complex system, and investigate how these perspectives can be effectively combined and used within the dependability assessment of a specification. This activity will include investigating the relationship between the formal specification and the associated dependability viewpoints. The identification and integration of dependability perspectives may be considered for a number of application domains.

The devised dependability perspectives will be incorporated into the coherent specification methodology developed within work-stream one.

○   Dependability Within The Development Process

Having defined the dependability criteria within the specification, the purpose of this activity will include an investigation of the relationship between the development lifecycle (including the proof of safety invariants, refinement of the dependability criteria/perspectives and validation/verification approaches) and the dependability lifecycle which includes safety analysis (eg, the relationship between fault trees, proof of safety invariants, and static analysis tools), fault detection/protection and failure detection/containment.

This activity will primarily be aimed at considering where the dependability analysis model can be effectively supported by the developing formality within the development and V&V approaches.

Tool support is essential for the practical application of dependability assessment techniques and therefore throughout the analysis we intend to identify areas where we believe that further tool support could be effectively employed.

○   Analysis Of Software Root Failures

Further work will be undertaken to investigate the root causes of software failures together with other contributory factors. This analysis will contribute to the identification of necessary dependability perspectives.

SRD have developed techniques for such dependability analyses which classify dependent failures on operating plant; SRD have also developed computer software in the form of a database structure to store and further analyze the information. We aim to develop a similar classification scheme for software failures. SRD can then examine the information to identify trends and patterns.

The identification of root causes of software failures, as defined within the dependability analysis work, provides a strong influence on the work-stream two work development approach. Once root causes are identified it should be possible to add techniques/tools to the development approach to avoid such failures.

5.   Summary

SafeFM aims to provide guidance on the cost effective requirements capture, development and assessment of high integrity systems which will be responsive to the changing competitive environment in which we operate.

These guidelines, which will be based upon practical experience, aim to build upon and further improve currently available technology by selectively applying formal development techniques. In particular the project aims to identify how disparate requirements can be defined, these requirements developed effectively and the requirements capture and development assessed. SafeFM aims to effectively bring together, in the form of a guidance report, technology which is emerging within current national and international standards.

REFERENCES

[1]   Bjorner, D, A ProCos project description: ESPRIT BRA 3104, Bulletin of the EATCS, 1989, 39, pp. 60-73

[2]   UK Mod, The Procurement of Safety Critical Software in Defence Equipment (Part 1: Requirements, Part 2: Guidance). Interim Defence Standard 00-55, Issue 1, Ministry Of Defence, Directorate of Standardisation, Kentigern House, 65 Brown Street, Glasgow G2 8EX, Uk, 5 April 1991

[3]   Bradley PA, DARTS European Software Safety and Reliability Guidelines Initiative, in Reliability of Programmable Electronic Systems, SRD Association, AEA Technology, SRDA-R6

[4]   A.P Ravn, H Rischel, V Stavridou, Provably Correct Safety Critical Software, Procs of SafeComp 90, pp 13-18, BK Daniels, Ed, Pergamon Press, 1990

[5]   Wensley, J et al, SIFT: design and analysis of a fault-tolerant computer for aircraft control, Proc. IEEE, 1978, 60, (10), pp. 1240-1254

[6]   Zhou C, Hoare CAR, Ravn AP: A calculus of durations, Information processing letters, 1991, 40, (5), pp. 269-276

[7]   P Mukherjee, V Stavridou, NewThink: An Orwellian Specification Language for real-time safety critical systems, Procs of the 4th Euromion Workshop on Real-Time Systems, pp 128-135, IEEE Computer Society Press, June 1992

[8]   Morgan C, Programming from Specifications, Prentice Hall, 1990

TOOL SUPPORT FOR AN APPLICATION-SPECIFIC LANGUAGE
(DTI/SERC Project ITD4/1/9035)

R.E.B.Barnard

GEC ALSTHOM Signalling Ltd
Manchester, England

1.   INTRODUCTION

This paper describes work being carried out collaboratively by GEC ALSTHOM Signalling Ltd, Westinghouse Signals Ltd, and the University of Manchester Institute of Science and Technology, with the support of DTI and SERC, under their Advanced Technology Programme Safety Critical Systems initiative. The work covers research into the design and checking process for data written in application-specific languages.

The project concentrates on one particular language - the well-established SSI data language - which is used very widely to define the functionality of safety-critical railway signalling interlocking installations based on the "Solid State Interlocking" (SSI) system.

2.   THE SSI SYSTEM

2.1.   Hardware

Each Solid State Interlocking installation has the function of controlling the points and signals on one section of a railway network. Interlocking between these controls ensures that safety of train movements is maintained at all times.

The overall concept of SSI [Cribbens87] is quite simple to understand, and may be summarised as follows (See Fig 1.):

- A centralised safety-critical microcomputer system issues commands to, and receives status indications from, up to 63 safety-critical controllers distributed along the railway.

- Each trackside controller contains a dual-channel microcomputer system, for safety purposes, and it drives and monitors either one or two colour-light signals, or up to four sets of points, as well as monitoring the status of train detection systems, etc..

- In the interests of availability, the centralised equipment is fault-tolerant, with active redundant 1-out-of-2 and 2-out-of-3 hardware configurations used for non-safety and safety portions of the system respectively.

- Communication between the centralised and distributed parts of the system is achieved by securely-coded digital data links, duplicated for availability, carried on screened twisted-pair copper cables for distances of up to 40km. In addition, 64 kbit/s channels in optical fibre PCM systems may be used for longer distance communication.

- The detailed functional requirements of each installation are defined by the applications data, which is compiled into about 60k bytes of object data, and stored in EPROMs in each channel of the central equipment. Data editing, compilation and simulation are carried out off-line on multi-user Design Workstation Systems (DWS), using software developed specifically for SSI.

The SSI system was developed collaboratively by British Rail, GEC ALSTHOM and Westinghouse Signals Ltd, beginning around 1980, with a first installation entering service in September 1985. Since that time, incremental improvement of the SSI system has taken place, with a large number of completed installations and orders in hand for the UK and overseas, including applications on British Rail, Docklands Light Railway, Manchester Metrolink, as well as in France, Belgium, Korea, Australia, Hong Kong, India, etc.

The SSI system is now specified for *all* new signalling projects on British Rail, having replaced the relay-based systems previously used [IRSE91]. It offers cost savings over the earlier systems, as well as permitting more testing work to be carried out in the design office, thus reducing the time needed on site. Since the introduction of SSI, it has been possible to reduce project timescales significantly. The existing design office staff have in general been retrained to prepare and check the data, instead of designing relay circuits, and they are still able to utilise their specialised signalling knowledge in this new role.

2.2.   The SSI Data Language

For every new application of SSI, data must be prepared for the interlocking system, the technicians maintenance terminal, and the DWS simulator system. The data input to the DWS computer is in the form of a series of source files, which are then passed through several compiler programs to create the sets of object files used to program EPROMs for each type of hardware module. Once these data EPROMs are installed in the hardware, the fixed interlocking program uses the SSI data to define the conditions to be applied in each stage of interlocking operation.

The SSI source data files include declarations of signalling functions, by type. These include:

- Track circuits (to determine occupancy of the layout by trains), signals, points

- Routes (permitted paths from signal to signal through the layout)

- Flags (logical functions defined by the applications engineer), timers (used to delay interlocking operation, e.g. to give trains time to stop)

- Route requests, panel inputs and indications (non-safety interface with the signalman)

As well as these purely declarative files, there are data files which include the logical conditions which must be applied in various stages of interlocking operation. These include:

- Input and output telegrams (defining the logical relationship between bits in the telegrams sent to, and received from, trackside controllers and signalling functions within the interlocking)

- Flag operations (logical conditions for releasing routes, and for implementing user-defined functions)

- Panel requests (logic associated with the operator interface)

- Points free to move (conditions affecting point operation)

- Map data (a simplified map of the layout, used by the interlocking program to search through the layout)

The interlocking program accesses some of these files regularly, others as required. The data statements are punctuated to guide the interlocking program, and are laid out to particular standards to aid the manual checking process.

3.   THE "SSI TOOLS" PROJECT

3.1.   Background

There is a large investment of skilled resources being made today by almost every railway authority and specialised railway signalling contractor, in the design and checking of application-specific data for use by the signalling systems controlling each section of the railway network. This data is safety-critical in nature, defining as it does the function of the signalling system. This task has to be carried out, in some form, for every computer-based signalling system, whether it is safety data for interlocking or Automatic Train Protection (ATP), or non-safety data for control centres and other functions.

This need for skilled resources is increasing, partly due to the increased amount of resignalling work needing to be done (and made more affordable by the reduced cost of programmable electronic systems), and partly as a result of increasing concern about the need for thoroughness in safety-critical checking and testing work. The widespread introduction to use of systems such as Automatic Train Protection will produce the requirement for vastly more safety-critical data in the future. The difficulty that has been experienced, by railway authorities and contractors alike, in training these skilled staff limits the amount of signalling work than can be undertaken at present.

Automation of the data design and checking process, which could enable the skilled resources to be concentrated in certain aspects of the work, is therefore a very attractive proposition, potentially offering significant returns on investment both directly, due to cost savings in signalling work, and indirectly as a result of improved performance of a more modern railway.

3.2. Project Organisation

The SSI TOOLS project was conceived by GEC ALSTHOM Signalling Ltd, and Westinghouse Signals Ltd, to investigate ways to overcome the constraint on signalling work that exists at present. Both companies are large users of the SSI Design Workstation System (DWS) for entering, editing, simulating and compiling SSI data.

The University of Manchester Institute of Science and Technology (UMIST) have been involved with GEC ALSTHOM Signalling Ltd in research into the use of knowledge-based systems to assist in the generation of data for SSI systems, and therefore they were a logical partner in the SSI TOOLS work, to provide an injection of mathematical and specialised reasoning experience to complement the signalling and software experience of the industrial partners [Palmer92].

British Rail are not able to take part in DTI-sponsored research work themselves, but are clearly very interested in this area. They have research work under way in complementary areas, and it is hoped that they will co-operate actively with the SSI TOOLS project team, by exchanging ideas, reviewing each others work, etc.

An initial project proposal was submitted to the DTI in January 1991, and a revised submission was made in July 1991. The offer of a grant was received from DTI in January 1992, and following final agreements being signed, the project began on 1st May 1992. The planned project duration is three years, with resources averaging two engineers full-time from each of the industrial partners, drawn from their signalling, software and safety specialists, together with a full-time research assistant from UMIST.

The SSI TOOLS project involves technical work in four main areas, as follows:

- Abstraction of railway layout details from signalling plans

- Automated data generation from layout specifications

- Hazard analysis of railway operations, leading to automated generation of sets of safety requirements for particular layouts

- Assisted proof of compliance of data with these safety requirements.

Detailed responsibility for the technical aspects of this work has been distributed to the three partners, although provision has been made for active sharing of work in areas where mutual benefits will result, e.g. in assimilation of experience in the use of formal logics and reasoning strategies, where the experience of UMIST can usefully be transferred to the industrial partners to help them to gain the experience needed to use such techniques in their future products and systems.

3.3. Technical Objectives

The overall objectives for the project include the development and demonstration of a set of prototype tools for designing and checking the data for signalling installations. To limit the scope of the project to manageable proportions, only signalling principles currently in use on British Rail are being considered. As the project progresses, it is recognised that it may be necessary to further restrict the scope of work, to ensure that a complete set of tools working together can be demonstrated.

3.4. Safety Philosophy

Since the data for SSI interlockings is directly safety-critical (a single error or omission could cause a collision or derailment), the safety philosophy of any design method must be carefully studied. There is no point in researching an approach which will ultimately lead to the need for highly complex software tools, whose safety integrity level [IEC91] must be very high, since the cost of designing and maintaining this software is likely to be unreasonable.

Several observations can be made, as a result of our work on the safety philosophy of the SSI TOOLS project:

- There is only one wholly definitive "correct" representation of the railway, and that is the real railway itself. All of the many overlapping representations are, to a greater or lesser extent, imperfect simplifications of reality - which is of course immensely complex.

- Assurance of safety always ultimately derives from diversity, and the probability that errors in one process will be revealed by comparison with another process. This diversity may take many forms. For example, information produced by one person may be checked by another, or it may be possible to show that two different representations of a situation are equivalent.

- The more complex a representation is, the more scope there is in it for errors, and the more time-consuming it is to check.

In analysing the sources of information used in the signalling design and checking process, it is necessary to consider their likely accuracy, and the means by which errors in them would be detected.

The safety philosophy adopted in the SSI TOOLS project may be summarised as follows (See Fig 2.):

- To derive and check a minimal representation of the actual railway layout from various more complex sources, which must each be considered to contain errors (e.g. CAD plans of the railway, specifications, etc.)

181

- To use this minimal representation of the layout in two very different processes - a design tool chain, implementing the Codes of Practice for designing SSI data, and a safety checking tool chain, implementing the signalling principles shown to be necessary to protect against defined hazards in railway operation.

- To prove the equivalence of the output of the design tool chain, which is the SSI data to be used in the actual installation, with the output of the checking tool chain.

4.   PROJECT WORK

The following sections give more detail of the approaches being adopted in the course of the work in the four areas noted above. The relationship between the four activities, and between the various tools being developed, is shown in Fig 3.

4.1.   Representation of Railway Layouts

Underpinning all the other activities in the SSI TOOLS project is the need to have a machine-readable description of the railway layout to be signalled.   A prime requirement for any such description is the need to elaborate it later, for instance as additional functions such as Automatic Train Protection are required. A number of alternative representations were considered for this task, including:

- Tabular representations.   Such representations are used at present to define formally the required function of signalling systems. These tables (known as Control Tables) are prone to errors and omissions, and are difficult to check. For the research project, a more accessible representation was sought.

- Minimal schematics. For most purposes, signal engineers find it easy to work with schematic representations of the actual track layout, with little-needed information (such as geographical features, exact geometric track alignments, etc.) excluded.

- Ordered lists of elements. These can be used to describe layouts, with the order of the elements defining the connectivity, and with cross-references from one list to another defining the actual track topography.

- Object-oriented decompositions.   In these approaches, the railway layout is considered as being made up from instances of various kinds of objects, each having a defined connectivity with other objects, as well as having attributes, and behaviours. This approach has the advantage that additional objects or additional attributes can be added easily without destroying the compatibility with earlier versions.

The preferred approach to railway layout representation is a broadly object-oriented decomposition, with additional knowledge structuring based on an overall Railway System Model. The purpose of the system model is to further partition and filter the large variety of information typically present within layout schematics, according to its usage in the signalling design process. This primarily involves identifying how the non-signalling and physical world objects within the area of the layout constrain and modify the otherwise idealised functionality of the higher levels of the signalling object class hierarchy. As a simple example of this, level crossings (which provide no benefit to the operation of railways as such) require specialised signalling system functionality, as well as additional lineside equipment. The system model is outlined in Fig 4.

The overall objective of this phase of the project is to write a specification for layout descriptions, and then to produce a tool (the Layout Specification Generator (LSG) - in effect a prototype user interface) to enable such layout description files to be produced for the test layouts to be studied during the project. The display of these layout description files in a form which can be checked is very important and, for this purpose, a minimal schematic is probably ideal.

It is accepted that, nowadays, most of the information about railway schemes is stored on large design office CAD systems, and therefore the eventual LSG tool may obtain some or all of its information from this source. The objectives at present are concentrated on understanding the process and the various sources of knowledge about the application.

4.2. Hazard Analysis

Surprisingly little work seems to have been carried out by railway authorities on the "top-down" categorisation of hazards and the allocation of responsibility for providing defences against those hazards to particular subsystems. Therefore, a strategy was developed for the production of a structured list of hazards to railway operation, and the identification of those which may be controlled by the signalling system. It is intended to obtain feedback from railway authorities about the validity and completeness of this hazard analysis work.

Historical data, the intuition of experts, the analytical skills of specialists, and "brainstorming" techniques, have all been used to review and challenge the initial list of hazards. The list is categorised to determine which hazards fall within the compass of the signalling system.

This work leads directly on to a more closely focused Safety Analysis related to national regulations, British Rail standard signalling principles, etc. This work permits a list of the properties of an individual signalling system needed to ensure safety to be derived from the track layout.

The final stage of this part of the project consists of the development of a prototype tool to derive sets of safety properties for actual layouts from files produced by the LSG tool mentioned above. This Safety Requirements Generator (SRG) will provide one of the inputs to the tool which will be used to prove the correctness of SSI data for a particular scheme.

4.3. Automatic Data Preparation

GEC ALSTHOM Signalling Ltd and UMIST have been carrying out research into the task of automatically generating SSI data directly from track layouts, under a previous SERC-supported project [Palmer92]. This earlier work was one of the enabling activities that make the present project feasible. One of the conclusions of this work was that, to achieve the maximum benefit from the automated production of application-specific data, the tools used to define the track layout, generate the data, and prove the correctness of the results, must all be closely integrated. A key element of this integration is the layout definition specification referred to above.

As part of the present project, the earlier work on Automatic Data Preparation (ADP) using knowledge-based techniques will be reshaped, so that the layouts are defined in the standardised file format. Much of the earlier research work will be incorporated into a new ADP tool.

One of the most interesting aspects of the ADP work is in the partitioning of knowledge and rules about the process into different areas of the tool, e.g.:

- Knowledge about track layouts, which would change if the form of the user input interface were to be changed in the future, leaving the other areas of knowledge essentially unchanged.

- Knowledge about the SSI technology, which would be changed as the SSI system itself were to be developed or replaced, or adapted to suit other railways.

- Knowledge about railway signalling, which would consist of two parts, one essentially fixed for all railway authorities, and the other highly dependent on the rules of each railway authority. This latter portion would be changed each time the tool was used for SSI for a different railway.

Each of these knowledge bases is structured in a form which, as far as possible, is understandable to experts in the particular domain. Of particular relevance to the later exploitation of the SSI TOOLS project is the fact that "signalling" knowledge is contained largely in one part of the system.

4.4. Proof Checking

It is a primary requirement of the project to provide a means by which safety properties of layouts may be validated against the SSI data.

The proof checking part of the project must result in the development of a prototype proof checking tool, or Proof Assistant (PA). This tool is intended to permit comparison between the SSI source data generated by the ADP tool, and the set of safety requirements derived from the layout files by the SRG tool. The PA tool will reason about the equivalence of these two different representations of the requirements, and will permit manual intervention, if required, as an aid to proof. An important requirement in this tool is to provide a traceable record of the proof

184

argument used, to provide an audit trail to be used as part of the safety engineering process.

The safety-critical nature of the SSI data means that this validation procedure should be implemented with a high degree of confidence. This entails the use of formal methods in the specification, design and implementation of the validation procedure, and the need for a sound mathematical model of our validation strategy. The intention is to use Z on this project.

An important consequence of the need to verify formally that the SSI data meets certain safety properties is that it must be possible to specify the SSI data statements themselves in a formal manner. This is an area into which British Rail have recently put some effort, in this case using VDM notation. The results of this BR work are available to the partners in the SSI TOOLS project.

UMIST have investigated the applicability of several logical formalisms for reasoning with the SSI data. The formalism currently favoured has the following advantages:

- Validity checking for sizeable fragments of data may be implemented on a computer of practical capacity.

- It offers the possibility of producing a theorem prover rather than just a proof assistant

- In producing a theorem prover, we may avoid (or reduce) the need for users to have unnecessary specialised knowledge.

- Our logical formalism (and thus the tools that are to be produced using it) is potentially applicable to areas other than the particular one currently under investigation.

5.   EXPLOITATION OF THE RESEARCH

The two industrial partners in the project both have a continuing need to prepare and check large amounts of SSI data to support their present business. It is possible that the demonstration tools produced in the course of the project will be usable under controlled conditions within the participants organisations, to give some direct benefit in this process.

What is more likely is that further development of these tools by the partners will lead to the production of a suite of tools that can be widely used in the future, for work involving the SSI data language (i.e. the present generation of equipment or its data-compatible successors). These tools will be used by the following types of organisation:

- Signalling contractors in the UK and overseas, who design SSI signalling systems

- Railway authorities in the UK and overseas who use SSI systems - if necessary with adaptation of the toolset to suit the different signalling rules and practices on other railways

But perhaps a more significant benefit will arise from this work. As a result of the study of the design process for such a data-driven system, the designers of future systems will have a much clearer idea of the requirements that widespread application put on the conceptual system design. They will be able to design in the tools, methods and practices from the outset; no doubt leading to more efficient and demonstrably safe application practices for future generations of equipment.

There is also much scope for application of these techniques in other industries where data-driven systems have to be configured to the requirements of a particular application.

There is no particular reason why the use of such design and checking techniques need be restricted to safety-critical systems; the task of configuring non-safety control and information systems in railways and in other industries involves similar concern about productivity and accuracy of design and checking resources, and the need to de-skill certain tasks. The insights given by this research will, it is hoped, assist in the design of "easy-to-apply" technology in those areas as well.

6.   CONCLUSIONS

The partners in the SSI TOOLS project are applying a great deal of analytical thought to a safety-critical design process, which is known to be susceptible to human error. This is perhaps not a common activity, certainly in industry.

Our conclusions so far are that the insights that this work is giving us into the nature of the design process are certain to be valuable, both in the short term, with better productivity of design staff and reduced project timescales, but also in the longer term, in the design of future generations of data-driven systems.

Acknowledgements

Thanks are due to GEC ALSTHOM Signalling Ltd, Westinghouse Signals Ltd and UMIST, for agreeing to the submission of this paper, and to all the members of the SSI TOOLS project team for their advice on how best to describe a research activity that is still in progress.

References

[Cribbens87]   Cribbens A H: "Solid State Interlocking (SSI): an Integrated Electronic Signalling System for Mainline Railways". Proc IEE Vol. 134 Part B No 3, May 1987, p.148.

[IRSE91]   Institution of Railway Signal Engineers/ed. Leach: "Railway Control Systems". A & C Black, 1991.

[Palmer92]   Palmer J W and Renfrew A C: "Issue raised in Development of a large Knowledge-based system for the design of Railway Signalling Schemes". IEE Intelligent Systems Engineering Conference, Edinburgh, August 1992.

[IEC91]   International Electrotechnical Commission: "Software for Computers in the Application of Industrial Safety-related Systems", IEC 65A(Secretariat)122, Version 1.0, 1st August 1991.

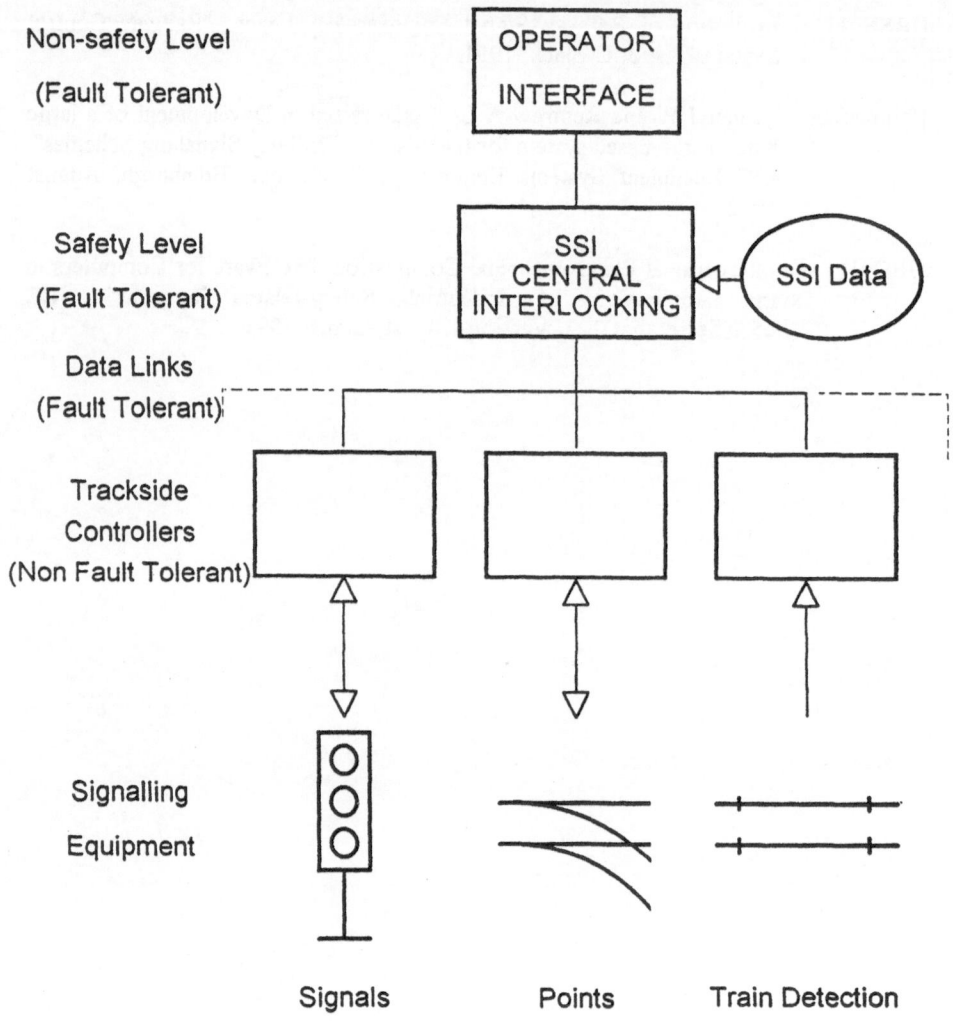

Fig 1. The SSI System Concept

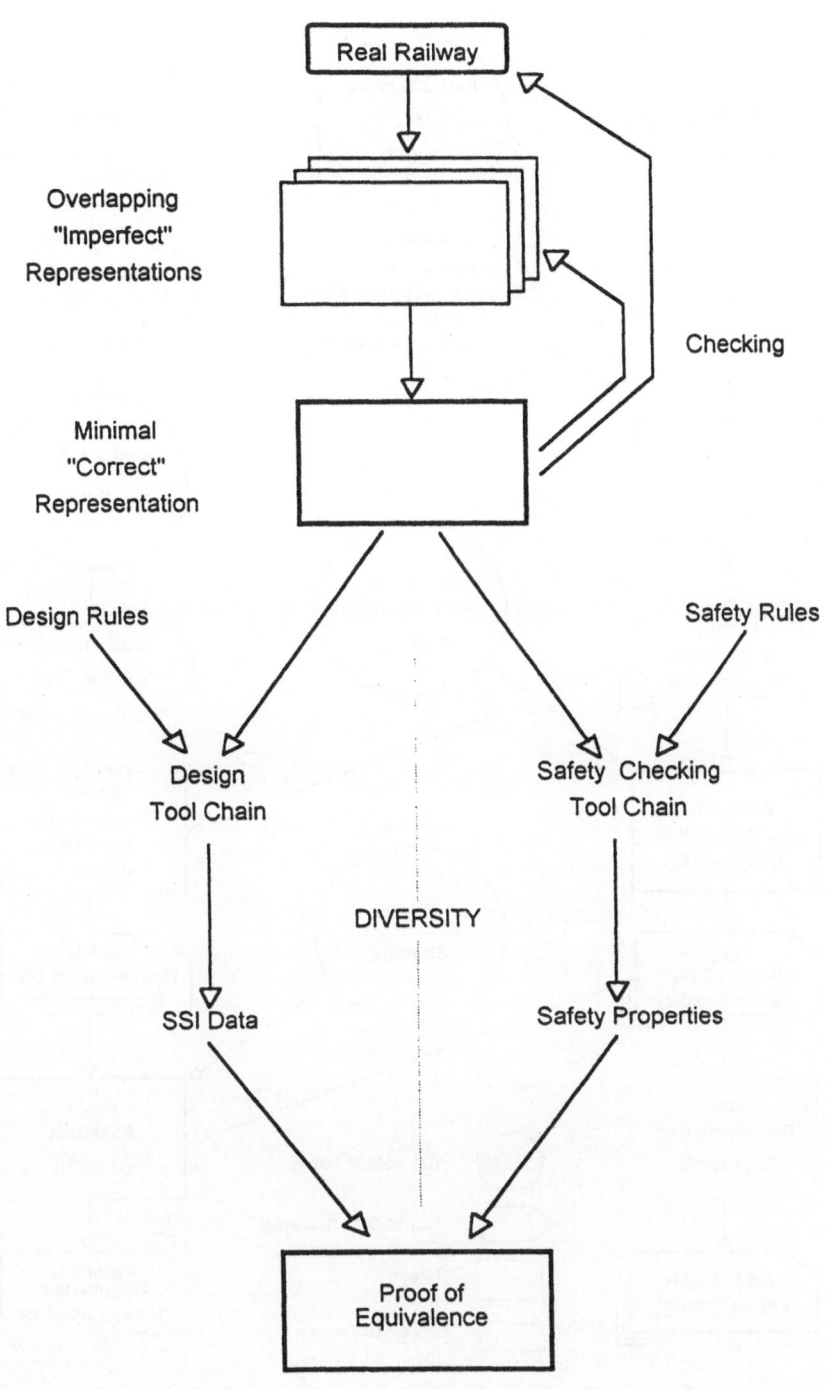

Fig 2. Safety Philosophy

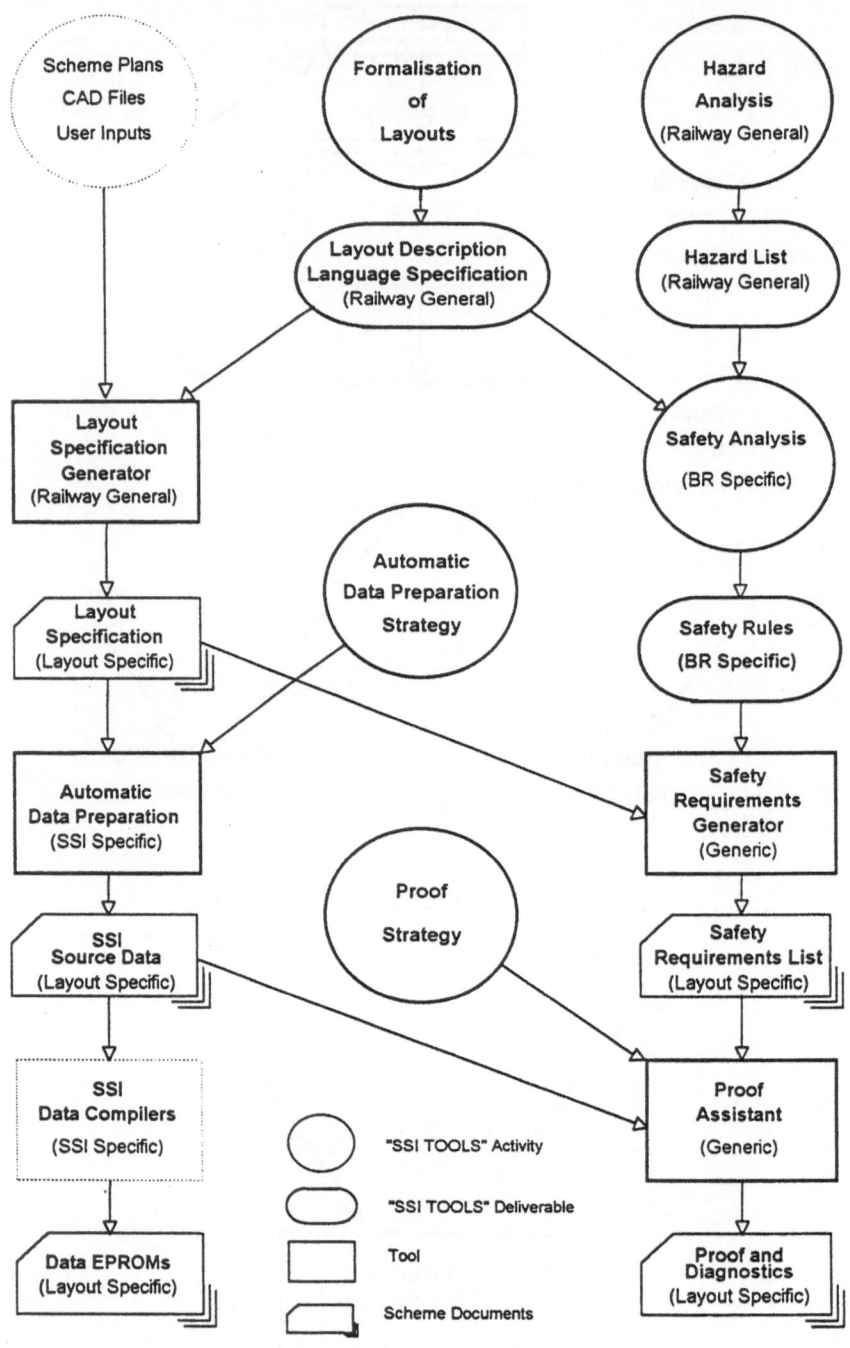

Fig 3. Scope of "SSI TOOLS" Project

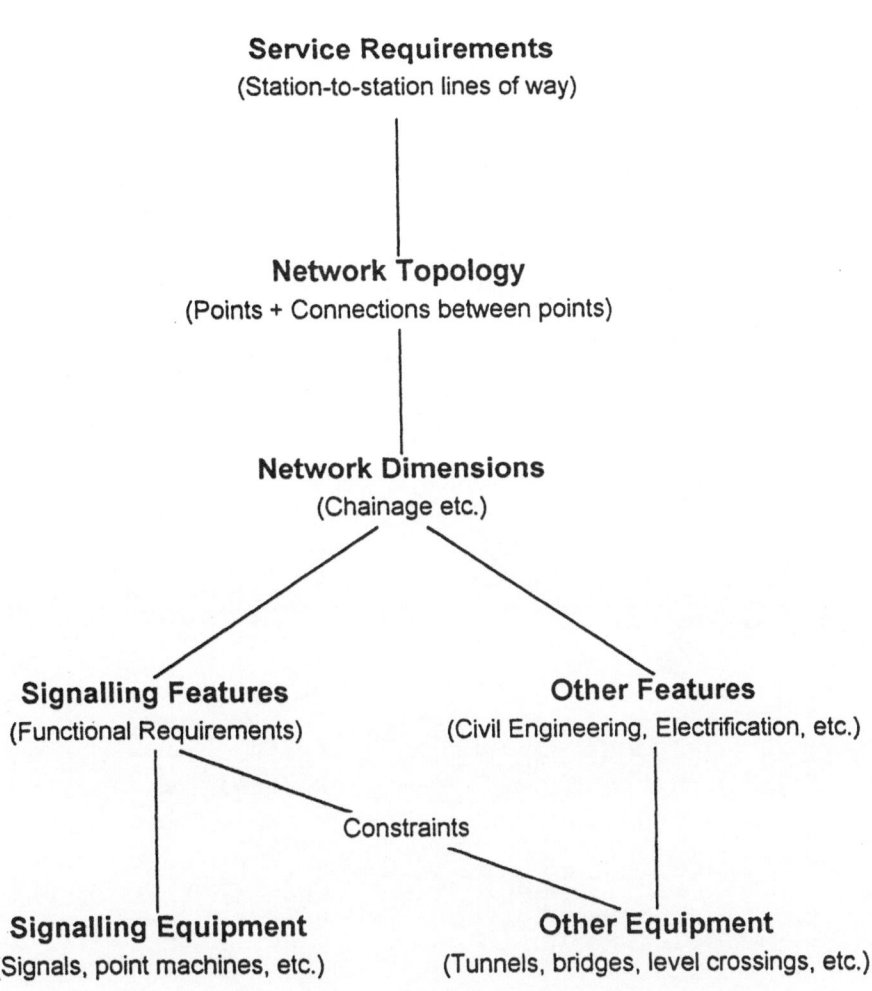

Fig 4. Layout Representation Hierarchy

PART III

ACHIEVING AND EVALUATING SAFETY

Thursday 11th February 1993

SAM—A Tool to Support the Construction, Review and Evolution of Safety Arguments

Justin Forder
Logica Cambridge Ltd.
London, UK

Chris Higgins
University of York
York, UK

John McDermid
University of York
York, UK

Graham Storrs
Logica Cambridge Ltd.
Cambridge, UK

Abstract

As pressure grows for stricter controls over the design and deployment of safety critical systems, the need for organisations to produce safety cases also increases. This paper describes the work of the ASAM ("A Safety Argument Manager") project—a project which has been concerned with providing tool support for the construction, review and evolution of safety cases. First, we present our view of safety and the issues involved in designing safety-critical systems, to explain the rôle of the safety case. Then we describe the project's objectives and set out the basic conceptual framework for safety cases developed by the project. After a discussion of the notation and style of argumentation that we have adopted, our prototype software tool, SAM, is described in some detail. The final section gives conclusions and directions for future work.

1   Introduction

There is growing public awareness of safety issues, and of the increasing rate at which complex technological artefacts which can affect safety are being developed and deployed.   This awareness is, at least in part, stimulated by recent accidents, e.g. Piper Alpha and the spate of aircraft crashes in the last few years. With growing awareness, there is increased pressure from the public, professional societies and some certification agencies for the development of stricter controls over the design and deployment of safety critical systems. Whilst many approaches to this problem are advocated, most of them involve the production of a safety case—documented reasons why a system is believed to be fit to deploy.

Experience suggests that developing such safety cases is, in itself, a difficult task. This paper describes the work of the ASAM ("A Safety Argument Manager") project which aims to support the process of developing safety cases through the provision of an improved conceptual framework for analysing safety cases, and the implementation of a support tool, known as SAM ("Safety Argument Manager"), based on those concepts. Before we discuss the project, we briefly consider what is meant by safety.

1.1 The Nature of Safety and the Design of Safety Critical Systems

There are many definitions of safety, and sometimes a distinction is made between systems being safety critical and safety related, dependent on the degree of harm they can cause. We take the view that safety is concerned with *absolute* harm, that is irremediable or irrecoverable damage. The damage can be to individuals, to property, or to the environment. Safety is a systems issue. Computer systems, and hence software, can only influence safety if they are used to control some physical process which can lead to harm. Thus, although we wish to build computer based tools to support safety analysis, our aim is to support safety cases about systems implemented in a mixture of technologies, and involving humans, not simply implemented as computer systems or in software.

Safety is not an absolute issue, but a relative one. That is, there is no real notion of an 'absolutely safe' system, and we need to establish the level of risk we are willing to tolerate when using a system. Nonetheless we can draw an interesting distinction:

- intrinsic safety—a system is intrinsically safe when there is no possibility of it causing, or failing to prevent[1], absolute harm;

- engineered safety—this term applies when a system has been designed to minimise risk, or reduce it to an acceptable level.

There are relatively few, if any, intrinsically safe systems, although some systems fall into this class if we ignore misuse. In most cases of interest we have to engineer systems to achieve 'acceptable levels' of safety. This is intrinsically risk-related as a system is 'more acceptable' the lower the level of risk, although acceptability is often influenced by the benefit of using the system, so the analysis is not straightforward. The development process and assessment (safety analysis) is driven by the risks and the design of mechanisms to combat the risks.

The design process starts by identifying the hazards of using the system. For example, with an anti-lock braking system (ABS) one hazard is failing to provide braking at speed. Having identified hazards we establish risks (the failure probability multiplied by the severity, e.g. the likely loss of life if the hazard arises[2]) and engineer the system so that the

---

[1] This phrase is required as some systems are purely preventative measures, e.g. reactor protection systems.

[2] Analysis of risk is quite subtle, and this simple definition is not always adequate, but it will serve for our purposes here.

risk associated with all hazards is acceptable. What is deemed acceptable depends on context and public perception of risk [Jones-Lee 87].

There are two major strands to the development process: the design and the safety analysis. Although these are often thought of as separate, they are intrinsically linked. The safety analysis is concerned with identifying the ways in which the system can contribute to the hazards, either through normal operation or through its failure, and quantifying the risks, so far as possible. The design is concerned, *inter alia*, with introducing mechanisms, often referred to as safety features, for reducing the risk to an acceptable level. If the safety analysis shows unacceptable levels of risk then this leads to redesign, so the two activities are linked in an iterative process.

The design leads to the production of the system and associated documentation. The safety analysis documents the safety characteristics of the system. This enables us to identify the role of the safety case.

1.2 The Safety Case

The *safety case* is the documentation of the reasons *why system is believed to be safe to be deployed* and it reflects the design and assessment (analysis) work carried out in the development process. In many situations the safety case (or a summary thereof) will be the major deliverable to *certification*, that is formal approval by some appropriate authority that a system is fit to be used[3]. The safety case should contain a record of all hazards, known as a *hazard log*, and reasoned arguments, based on the safety analysis, why the anticipated operation of the system, given the anticipated inputs and failures, will not lead to the identified hazards. In practice these arguments will reflect design information, reliability calculations, safety analyses, and perhaps proofs of program properties [McDermid 92]. The hazard log acts as the index as it should be possible to find the safeguards against each particular hazard (safety features), and the reasons for believing them to be effective, by tracing from the hazard.

Typically safety cases are very extensive, reflecting the detailed analysis of the failure behaviour of the system, e.g. down to the level of individual mechanical or electronic components. There are many standard safety analyses, e.g. Fault-Tree Analyses (FTAs) which identify all the failures, and combinations of failures, which can (with a significant probability) give rise to a particular hazard. A typical safety case will contain such analyses, but will also contain much informal material, such as reasoned arguments why it is unnecessary to consider the effect of computer hardware failures on software. Existing safety analysis tools tend to support the more "formal" analyses, and not the more informal arguments—although these are often very important, reflecting basic principles and assumptions behind the safety cases which, if they prove unfounded, could result in much of the case being ill-founded. This identifies one of the main aims of the project—to provide better support for these informal aspects of safety cases.

---

[3] For example the airworthiness authorities such as the CAA are responsible for the certification of civil aircraft for use in the UK (in conjunction with other authorities).

1.3   Current Safety Cases

An assessment of current standards and industrial practices suggests that the quality of the safety cases presented could be improved by means of automated support. Most safety cases are developed using simple tools and databases, together with word processing facilities, and rely heavily on manual review of the completeness and consistency of documents. Whilst automation is not a panacea, and there will always be a need for human judgement in safety analysis, the consistency and visibility of safety cases could be improved by appropriate tool support—especially that which makes the informal aspects of the safety case explicit. In addition the provision of automation gives the possibility of reasoning about and analysing safety in ways which are not practical when all the relevant information is presented in paper documents. We now briefly summarise our perception of the weaknesses of current documentary safety cases as a precursor to setting out more fully requirements for SAM.

In almost all cases achievement and assessment of safety requires a multi-disciplinary approach, perhaps involving mechanical, electrical and software engineers and human factors experts, including psychologists. Some of the limitations of current approaches arise from the difficulties of identifying interactions between the concerns of different disciplines, although many arise from other causes such as system complexity and the difficulty of placing a boundary around the factors/issues relevant to the safety argument.

A major source of problems arises from assumptions that are made in system development and assessment. Many assumptions are made without explicit justification. If all justifications were stated explicitly then it would become possible to assess their validity. Perhaps a more serious problem is that many assumptions are never made explicit so their validity, and consistency with other assumptions, cannot be assessed. A computerised tool cannot automatically derive assumptions, but it can help to draw attention to the absence of assumptions and/or justification for assumptions. A practical example where rigorous justification of assumptions is required is in assessing independence of failure modes in redundant (replicated) systems.

A strongly related issue is completeness. Many arguments are incomplete in that some of the stages of the argument structure, like steps in a mathematical proof, are missing. These omissions may, in themselves, reflect implicit assumptions. Whilst the notion of completeness is somewhat subtle and, arguably, it is not possible to check completeness automatically it is possible to detect some omissions from arguments with automated support. Incompleteness is a significant problem with practical arguments about safety. Substantial errors have apparently gone unnoticed into operational systems [Leveson 86]. Automated support for argumentation gives the opportunity to check at least some aspects of safety case consistency. This notion of consistency extends to dealing with problems of change, and ensuring that the safety case and design stay "in step".

There are many examples where failure, or even the expected behaviour, of some part of a system or its environment has caused an unexpected failure in another part of the system. In many cases these problems arise because experts in one discipline fail to take into account (are unaware of) important characteristics of the system on which they are working. For example it is reputed that recent problems with Audis in the USA were caused by electromagnetic radiation from CB radios interfering with the operation of the computers controlling the cars' transmission. Again automation may enable cross-checks to be made between arguments relating to different disciplines, e.g. to see if all consequential failure modes have been taken into account.

Other difficulties simply reflect the scale of safety cases, or are exacerbated by scale. Typical safety cases run into thousands of pages. Thus it is essential to provide support for production of large safety cases, navigation through large cases, and change control in order to address industrial-scale problems.

None of the above problems can ever be completely avoided, mainly because they involve common-sense reasoning about the real world. However the intention is that SAM will alleviate the above problems by making safety arguments easier to inspect and by facilitating analysis of the validity of individual arguments and the inter-relationship between arguments.

1.4 The ASAM Project and the Scope of the Paper

The ASAM project was a collaboration between Logica Cambridge, the CAA and the University of York. It was part-funded by the DTI and SERC as part of the JFIT programme and followed earlier work, done by one of the partners, on supporting policy arguments in the Alvey DHSS Large Demonstrator Project [Storrs 91]. The primary aim of the ASAM project was to produce a "proof of concept" safety case tool, intended to overcome some of the difficulties outlined above, and focusing on the informal parts of the safety case. The project undertook work on two broad fronts: development of a conceptual framework for safety analysis, and provision of a series of prototype tools. The CAA, and to a lesser extent the other partners, carried out a number of trial applications of the tool in order to evaluate its capabilities and feed experience into the development of later prototypes. The paper describes all three aspects of the project's work.

Section 2 amplifies the requirements and objectives for the tool as agreed by the project. These requirements and objectives are presented at a fairly broad level—the analysis in the project was based on the consideration of a number of scenarios for use of the tool. Section 3 lays out the basic conceptual framework for safety analysis developed by the project and which, in many ways, is seen as the lasting value of the project. Section 4 discusses the functionality of the tool as it existed at the end of the project (October 1992). Finally, in section 5, we draw out some lessons about the strengths and weaknesses of the tool, and our intentions for follow-on development and exploitation.

2   Objectives and Operational Concepts

2.1   Objectives and High Level Requirements

The primary objective for SAM is that it should facilitate the production of complete system safety cases which are of higher quality than is normally achieved in current practice. These cases should focus mainly on informal arguments, but also show how standard safety analyses can be integrated into the overall safety case. The most important subsidiary objectives are:

• SAM should make safety arguments more visible, or inspectable, thus facilitating informed human assessment of argument quality, and hence improvement of quality (e.g. by rectifying omissions);

• SAM should support structuring of safety cases, in order to address (at least in part) the problems of scale;

• SAM should support models of the system design to support consistency checking between the design and the argument;

• SAM should automate some aspects of safety analysis, e.g. fault tree analyses or reliability calculations, to demonstrate that it can also support classical safety analysis techniques;

• SAM should have automated mechanisms for linking to other tools, thereby avoiding some sources of human error, i.e. SAM should have a role as an integrator;

• SAM should support the tracking of dependencies between parts of the safety case, facilitating the processes of inspection and change of safety cases.

These basic objectives lead on to a number of rather more detailed, but still high level, requirements. SAM should support:

• creation, modification, browsing and printing of arguments including the ability to follow complex argument chains;

• more sophisticated argument manipulations such as identifying the ramifications or consequences of change, and helping generate summaries of complex argument structures;

• the development, modification, printing and browsing of system models, e.g. architectural block diagrams, and linking these models to arguments;

• importation of supporting evidence from other tools, spanning the whole development process, to provide basic data on which the arguments depend;

- checking consistency of complex argument structures, and the relationships between arguments and models.

Whilst we could set out requirements in more detail, the above captures the notion of the basic functionality required of SAM which was used to drive the project.

2.2   Operational Concept

The aim in describing the operational concept for SAM is to indicate how we propose that the above requirements will be met, but from the point of view of the way in which the tool will appear to a user, rather than from the point of view of the internal system architecture. It is possible to think of SAM in a number of quite different, but compatible, ways. Consequently we briefly present three different perspectives on the system, each of which reflect different aspects of the tool's capability; the primary distinguishing feature of each form of the model is the aspect of (the documentation of) the safety case which it treats as being primary.

When using SAM in any given scenario (e.g. safety analysis during design, or analysis as part of the certification activity) we are interested in three different aspects of (the documentation of) a safety case, and the relationships between them. These aspects are:

- *documents*—printable/presented as hypertext, including diagrams, formal arguments, explanatory text, and so on;

- *models*—representing the system and its operational environment(s), and perhaps its development and evaluation environments;

- *arguments*—formalised representations of chains of reasoning giving at least the capability to follow chains of reasoning.

Because it is intended that SAM will be used in a number of different scenarios it is not possible (desirable) to constrain the order in which these three aspects of the safety case are developed. For example, in the design process, particularly in support of design decision making, it is likely that a system model will be produced first, followed by a set of arguments and a document. In the case of post-hoc systems analysis a document will probably already exist and actions will be required to extract arguments from the text, and little if any system model may be produced. Thus it is extremely important that these three components of a safety case can be developed in any order and can evolve together. There are three other key points.

First, typical safety cases are multi-disciplinary so few, if any, users of SAM will wish to see the whole safety case. Thus it is also necessary for SAM to support work through views, that is projections or subsets of the whole safety case.

Second, it is important that the documents produced by SAM can be of high quality—that is they read well and are technically coherent, rather than being a disjointed set of notes around the formal (analysable) argument content. Thus SAM should support the notion of "presentation views" of the safety case which correspond to coherent printed (or displayed) forms of the argument.

Third, although SAM is intended to be very much a "user aid" rather than a fully automatic tool, it must support some means of checking consistency of arguments, and between arguments and models to improve the quality of safety cases. Thus, for example, it should be possible to check that all relevant failure modes have been included in a failure modes and effects analysis, by checking the consistency of the arguments and the model.

Although we mentioned above the need to analyse existing documents and to represent them in a suitable internal form, SAM must be able to work with existing documents relating to a system of interest without being able to analyse them fully—it is clearly unreasonable to expect SAM to replicate the functionality of all relevant design and analysis tools. Thus SAM must be able to import documents and build relationships between those documents and the remainder of the safety case. It must also support change management where one version of an imported document is superseded by another and (ideally) change management should be supported at a finer grain size than the complete document. For similar reasons SAM must interface to other tools, e.g. simulators or program analysis tools.

The operational concept can be summed up by saying that there are multiple facets of a safety case and SAM must support work on these facets in any order, and give the user the ability to change between facets, translate between views of the safety cases, and support the development of high quality documents, relating the case to other system documentation in a traceable way.

SAM is intended to supplement human capabilities, not to supplant them—to help the developer of a safety case make the important decisions and arguments explicit and clear. Nonetheless some automated checking of consistency between the model and the arguments facilitates the production of a more "semantically rich" analysis tool. We can now set out the conceptual framework which underlies the tool, before setting out the capability of the tool produced by the end of the project.

3 The Conceptual Framework

One of the primary results of the project is the development of a sophisticated and extensive conceptual model for supporting argumentation in general, and safety cases in particular. This section summarises the key aspects of the conceptual basis for the tool, and also indicates some areas where further progress is required in developing an adequate conceptual structure for developing complex safety cases. The description also indicates what facilities are provided by SAM, although more detailed descriptions are provided in section 4.

3.1 Argument Representation—Toulmin Form

Toulmin form [Toulmin 58] [Toulmin 84] is a notation for expressing what Toulmin called "practical arguments". It is presented by Toulmin as being a development from the syllogism and is based on the assumption that each step in an argument—each *micro-argument*—can be expressed using a simple syntax. It will be useful briefly to review the reasons for the particular form that Toulmin devised.

If we make an assertion, Toulmin argued, we commit ourselves to the claim that it involves. If that claim is challenged, we will, normally, produce some facts to support it. The facts themselves may be challenged and, if so, will require their own support so that they may be agreed and the argument may progress. This gives us a basic and simple distinction between a Claim and the Data which support it. Of course, the Claim may still be challenged, not now on the basis of its factual support but on the question of why we believe those Data support that Claim. What the challenge demands are the "rules, principles, inference licences or what you will" that justify the step from Data to Claim. Such justifications Toulmin calls Warrants; "general, hypothetical statements which can act as bridges and authorise the particular step to which our argument commits us". The basic form of an argument is therefore:

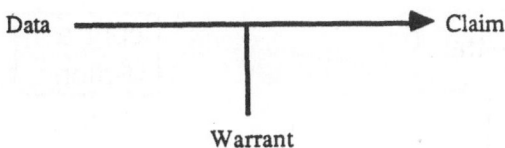

Figure 1: Toulmin's basic micro-argument form

The nature of our Warrant will affect the amount of force that the argument gives to the Claim. That is, we will need to qualify the Claim on the basis of the Warrant we have used. There are also conditions of exception or Rebuttal which indicate circumstances in which the general authority of the Warrant would not hold. What is more, the authority of the Warrant itself may be challenged, in which case we need to provide explicit support for it in the form of a Backing. A Backing is a statement of facts which, in the particular field of argument we are in, can safely be taken as certain or self evident.

Toulmin's complete argument form now looks like this:

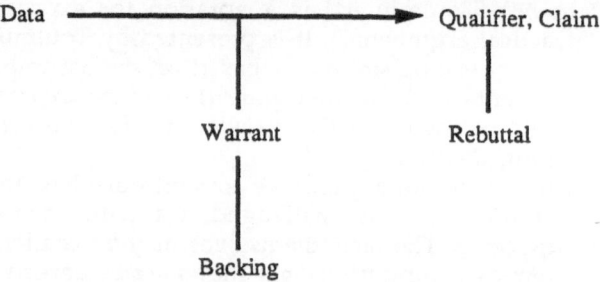

Figure 2: Toulmin's complete micro-argument form

For several reasons, we have chosen to extend Toulmin Form slightly and to give it a more precise interpretation than Toulmin himself has provided. Our Extended Toulmin Form notation (ETF) has a slightly different appearance from the original. Figure 3 shows the form currently supported by the ETF editor in SAM. The changes are mostly cosmetic (boxes around the argument components and the introduction of extra arrowheads on the lines) but there are others which are more significant.

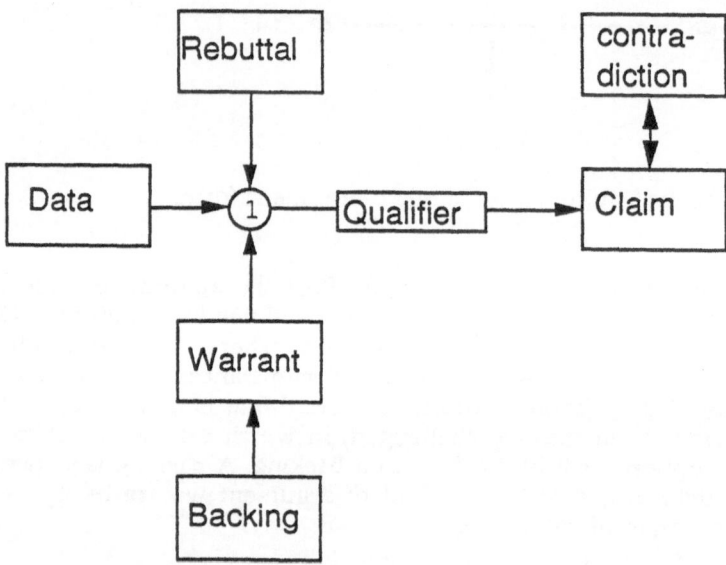

Figure 3: The SAM version of Extended Toulmin Form

We have added an extra node to the diagram which we call the Argument Node. This is the circle between the Data and Claim components. The Argument Node carries a unique label for the micro-argument (here shown as a number). This may be thought of as the *name*

of the micro-argument, which one could use to refer to the micro-argument as a whole rather than to any of its parts.

Another important change is that the Rebuttal is now linked to the Argument Node rather than to the Qualifier as Toulmin has it. It is to emphasise the fact that the Rebuttal is of the micro-argument's ability to make its Claim rather than of the Claim itself. To illustrate this, consider Toulmin's "Bermudan" argument [Toulmin 58]:

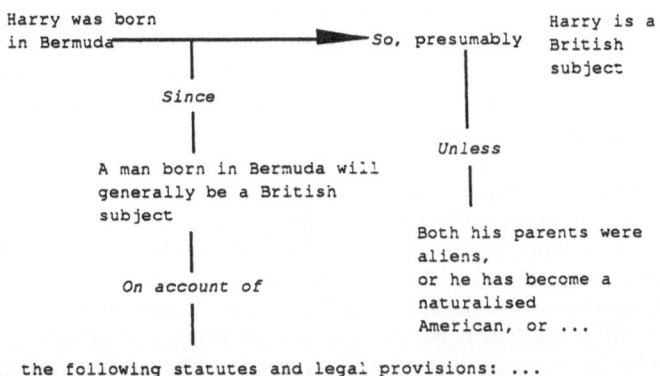

Figure 4: Toulmin's "Bermudan" argument

The implication from Toulmin's layout is clearly that the Rebuttal is a potential challenge to the *force* with which an argument can make its Claim, not of the Claim itself (a common misinterpretation). Hence its attachment by Toulmin to the Rebuttal. We wished to make it absolutely clear in our own layout, that the Rebuttal is of the Warrant's ability to sanction the move from Data to Claim. That is, it is a Rebuttal of the argument step itself rather than of the conclusion of the argument.

In either case, the Qualifier is not determined by the Rebuttal but by the degree to which we can reliably deduce the Claim from the Data on the basis of the Warrant. This means that the Qualifier too should be thought of as attaching to the Argument, as its effect on the Claim is only valid in the context of the micro-argument and does not propagate. We draw it attached between the Argument and claim partly for consistency with Toulmin but mostly to improve the readability of the micro-argument.

The final change is to add another new component to the form. This allows statements to stand in the rôle of Contradiction with respect to other statements. One of the most important uses for the new component is in the denial of Rebuttals but it also allows contradiction to be used structurally as part of the micro-argument network. Although the component is shown attached to the Claim in figure 3, it may in fact be attached to any component except the Qualifier.

One very important thing to note about our interpretation of ETF is that none of the components (except the Qualifier) are defined by the *type* of statement that they contain but by the *rôles* they play in a micro-argument. This is clearest in the simple case where the Claim of one micro-argument is also used as a Datum for another but it is also true in

205

such cases as the Warrant of one micro-argument being the Claim of another, or a Rebuttal also being used as a Datum, and so on. This means that, in analysing a written argument, it is impossible to tell the rôle that any particular statement will play in an argument without looking at the argument as a whole.

3.2   Safety Case Structuring

Practical arguments, such as safety cases, are intended to persuade, so they are structured in such a way as to maximise their persuasive effect on their particular audience. In making a safety case, the arguer employs a strategy for presenting his or her arguments and this strategy determines the rhetorical structure of the argument.

In our work we have presented the overall structure of an argument as a goal structure. The top-level goal of a safety argument might be to show that a system is acceptably safe to deploy. This goal is achieved (say) by showing that the system meets required standards for its configuration, its reliability and its quality (cf. the structure of the example given in [HSE 87]). Each of these goals is satisfied by sub-goals and so on until a complete lattice structure of goals has been created where the terminal goals can be satisfied directly by argument or by reference to the system model (see §3.3).

If we think of the reasoning steps that we describe in ETF as micro-arguments, we can regard the goal lattice as a *macro-argument*. It gives the overall structure of the argument and we can use it as the framework on which to attach strategies, contexts and solutions. The decision about how each goal in this structure is satisfied is a matter of strategy. This determines the sub-goals, if any, the contexts for the sub-goals and any assumptions that might need to be made in those contexts. The strategy may also describe the kind of argument to be used as a solution for the goal.

Figure 5 shows the relationship between the goal structure, detailed argumentation and model in the scheme we use on the ASAM project. The diagram is simplified in that, for clarity, it excludes details of meta-data and the fact that there may be many different "views" or abstractions of the model.

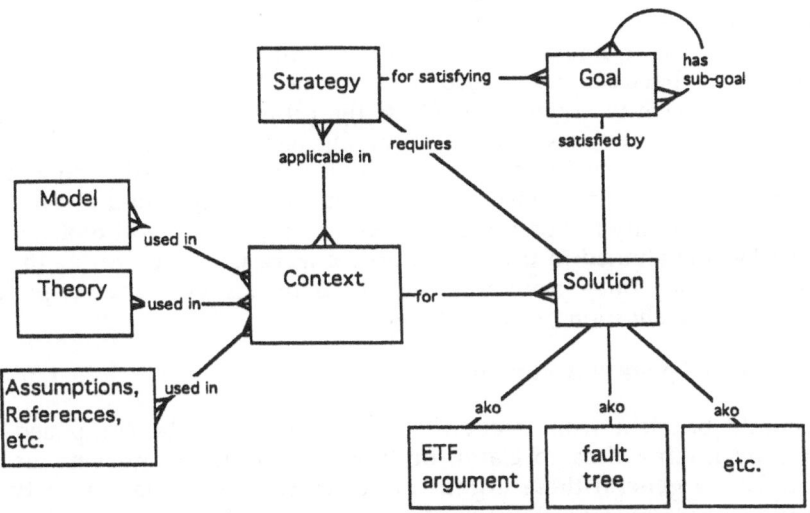

Figure 5: The relationships between the major conceptual entities in
ASAM.

The argumentation in a safety case is developed as the solution to a
goal. Goals may be, for instance, to show some aspect of a system is safe or
has some other property, or to obtain some data, or whatever. The
solution is the ETF argument, or the fault tree analysis, or the FMECA, or
whatever will satisfy the goal. Solutions must be developed within some
context and this comprises the system model, the theory for the model (i.e.
the rules or physical laws that are appropriate to the statements in the
model), assumptions that the arguer makes in the course of developing
the argument and references to data outside the current case. As interim
results are produced in solutions (e.g. when claims are produced in ETF
arguments) these too are added to the context.

3.3   System Model

In order to provide a basis against which to articulate safety arguments a
model of the system, its operational environment, and perhaps other
structures—e.g. the development organisation—is needed[4]. The model
can be provided at a number of levels of detail, or abstraction, ranging
from a very high level view of the system, down to a very detailed
engineering model, at the level of electronic components and their
interconnections. The former might be appropriate for a hazard analysis,
and the latter for an FMEA.

The production of a model is inevitably dependent on the system
under consideration, for example different issues are at stake when
considering an aircraft landing gear than when considering a nuclear

---

[4] The term *system* in system model is perhaps not very appropriate, but one that has been
employed throughout the project.

reactor shut-down (protection) system. Thus it is not possible to derive a single standard schema for the system model, but there is some commonality between applications of the tool. By taking an object-oriented approach to system modelling, the project aimed to provide a set of generally-useful sub-schemata (covering aspects of system structure, communications, interaction with the environment, hazards and failures etc.) which could be specialised further to meet the needs of a given safety case. This capability has only recently been added to the SAM tool, and has not yet been evaluated in practice. Further work is needed to see to what extent a schema can be produced which is useful across a range of projects, mainly by specialisation of the class hierarchy.

3.4   Chains, Nets and Summaries

As should be clear from section 3.1, arguments can be composed into chains by linking data to claim, or linking in other ways, e.g. through rebuttals. In general these argument structures form nets, and a typical safety case will contain many micro-arguments linked in a complex way. The topology of such structures is less important than the dynamics of their creation.

Simple chains may be formed in various ways, the most obvious being forward-chaining from primitive data to an overall goal, and backward-chaining. The choice between these two approaches will depend on the problem being solved, although backward-chaining from goals towards primitive facts in the model is perhaps the most likely. Other aspects of the structure may arise from challenges to the chain.

For example, a challenge may be made asserting that some rebuttal holds. The argument chain may then be extended to a net, by adding micro-arguments that show that the rebuttal does not hold. Other forms of challenge may be to the argument as a whole, leading to the provision of a warrant if none was present before, or to the provision of a micro-argument that supports the warrant. These micro-arguments may themselves be challenged, leading to further extension of the net.

Support for construction of argument nets is supported by SAM, and a graphical layout is used to simplify entry of the nets.

In developing large scale safety cases many detailed arguments are produced, and it is desirable to be able to summarise these. One simple example of the need to produce summaries is when a set of arguments based on a design of a system (in the model) at one level of detail needs to be abstracted to refer to the model of the design at a more abstract level. Also it is clear that the solution to a goal is often a summary of its immediate sub-goals. In some cases, e.g. when the summarising relates to some abstraction function in the design representation, it may be possible to generate summaries automatically. In general, however, this does not seem feasible as it is not possible for SAM to infer the purpose of the summary, and hence to determine what information to leave out. This issue received some attention in the project, but it is still an open question whether or not summaries can be generated automatically under some circumstances. At present the production of summaries is a manual exercise.

3.5   Formal Arguments

Extended Toulmin Form is a semi-formal notation for argumentation—it provides a formal structure relating informal statements. This gives complete freedom regarding the content of ETF arguments, but requires human reviewers to check the well-formedness of micro-arguments, and to check the consistency of statements different arguments with each other and with the system model.

By using a formal language for the statements in ETF, where appropriate, we give up some subtleties of interpretation, but gain the ability to do these checks mechanically. To allow this, SAM supports a typed first-order logic, the types, attributes and relations being given by the schema for the system model, and the individuals of these types being given by the system model itself. The user has a free choice as to whether a given solution should be expressed semi-formally or formally. If desired, both kinds of solution can be given for the same goal—the semi-formal solution possibly acting as a summary of the formal one.

3.6   Specialised Argument Forms

In the project's terminology, specialised argument forms are particular safety analyses, e.g. FMEAs, used to provide some data for a Toulmin argument, to provide data in the model, or perhaps to provide the solution to some goal. In general it is intended that high level arguments would be presented in Toulmin form and that supporting evidence would be presented in a specialised form. It was always intended that SAM would be a general tool, and support for specialised argument forms would be provided by interfacing to other tools, consistent with the notion of the separation between high level arguments and supporting evidence. However it is reasonable to use these specialised argument forms, on occasions, in high level arguments, e.g. fault trees can be used to model accident sequences.

In order to show that this principle could be supported, the conceptual model for SAM allows fault trees to be the solutions to goals (see fig. 5). No other forms of specialised arguments are currently supported directly by SAM.

4   The SAM Tool

4.1   The Development of The SAM Prototypes

The ASAM project developed a series of prototypes, beginning with SAM 0 in 1990 and ending with SAM 3.2 in October 1992. Macintosh computers with A3 screens were used, and implementation was done in Common Lisp and the Common Lisp Object System (CLOS) to maximise productivity. The prototyping approach was essentially incremental—each new prototype, generally speaking, had more functionality than its predecessor. Thus, SAM 0 was little more than a Toulmin form editor while the last prototype contains a (much enhanced) version of this editor as one of many other facilities. SAM provided progressively more functionality as our understanding of the problem and solution space grew, and its user interface evolved as we became more familiar with our users and their conceptual models.

The final SAM demonstrator, SAM 3.2, supports:

- construction, review and modification of a coarse-grained mainly-hierarchical structure for the overall safety case, and the use of this structure to navigate through the safety case content;

- construction, review and modification of fine-grained semi-formal and formal argument structures, within the overall safety case structure; and

- construction, review and modification of informal or formal models of the arguments' subject matter.

It allows annotation of content with reviewers' comments, records change histories of all elements of the safety case, indicates places in arguments which need review when changes are made, and supports consistency checking of formal arguments and models. Support for Fault Trees is included, as an example of a special-purpose notation.

The focus of the project has been to prove the concept that safety argumentation could be supported in this way. The project has therefore concentrated on including generic facilities in the SAM prototypes, and has avoided tackling problems which already have well understood solutions. We thus put considerable effort into developing facilities for structuring arguments, modelling systems and creating detailed argumentation in ETF notation. Other facilities, such as the hyperlink mechanism, change impact warning and the object-oriented database back-end, received less attention and some facilities, such as annotation, change history recording and fault trees, because they were well-understood in their own rights, were included in a fairly rudimentary form primarily to indicate how they would integrate with the rest of the system. Some problems, such as producing sensibly-structured linearised textual representations of a complete safety case, were also tackled only in a simplistic way, not because they were trivial but because they were too difficult and sufficient resource to tackle them was not available.

4.2 A Summary of The Facilities of SAM 3.2

SAM 3.2 was the last prototype to be produced before the end of the ASAM project. It is a large and facility-rich piece of software and the summary below, while covering most of the features it offers, cannot hope to cover everything. The summary is organised under a number of functional headings as an aid to understanding the overall structure of the system.

4.2.1 The safety case

Each safety case is held as a Macintosh document, which appears as a document icon in the Finder. The complete safety case may be held in the document file, or (in an experimental variant) the file may hold a reference to an object-oriented database on a networked server. SAM normally presents the document content to the user as a mixed text and graphics presentation in a number of windows. It is possible to ask for a linear text version of the document to be created, displayed in a separate, single window and printed.

Unlike earlier versions of SAM, which used a fixed layout of tiled windows, SAM 3.2 uses multiple independent windows to contain the various parts of a safety case. In order to make the management of these windows easier, a number of window management commands have been provided.

4.2.2 Argument structuring

The Goal Structure is a diagram showing a lattice of goals, each of which needs to be satisfied to complete the safety case and reach a decision about the safety of the system. It resides in its own window and can be edited using interaction styles common to all graphical editing in SAM. The status of the work on each individual goal (whether it is in progress, completed, or reviewed) is indicated in the diagram. If any of the work on a particular goal has unresolved comments, or needs to be reviewed as a result of changes, this is also shown.

4.2.3 The solution

The Solution is an argument expressed in one or more argument form(s) which satisfies its associated goal and is part of the larger argument which supports the ultimate claim about the safety of the associated system. It is assumed that for each goal only one solution in ETF argument form exists and it can exist in semi-formal or formal representation, or both. In cases where the solution appears in both representations, the two representations should supplement each other, rather than presenting two different arguments. The structures of the two representations need not necessarily mirror each other—one may be an elaboration or a summary of its counterpart.

4.2.4 System model

The Model is a description of the system (and its environment, etc.) that the safety case is about. SAM allows you to define multiple views of an (implicit) overall model; each of these views may be associated with one or more goals, and may be expressed informally as an imported diagram, or formally as an Entity-Relationship diagram or as text conforming to the formal statement syntax (see §3.5).

Before a model view can be used, a schema must be selected. Schemata and sub-schemata are created with a separate schema editor. Only one

211

schema can be used in a given safety case, although it may have many sub-schemata and there may be many model views.

4.2.5 Schema editing

The separate Schema Editor allows the user to define and edit entity types and relationship types, defining the attributes for each. Inheritance is supported, both for entity types and relationship types. The Schema Editor allows definition of the symbols and link styles associated with entity and relationship types, and includes a graphical model editor to let the user try out the results. A finished schema document is imported into SAM, where it can be instantiated to produce a system model.

4.2.6 Formal arguments and argument checking

There are two notations for formal models. One is the diagrammatic form which appears in model views. This is a restricted subset of a standard entity-relationship (E-R) model syntax. The other is a textual form, the syntax for which is discussed in §3.5.

The textual form can appear either in the textual view of a model view or as formal statements in ETF arguments in formal solutions. In the textual view of a model view, the formal statements cannot be edited. In formal ETF arguments they can be both entered and edited.

Normal ETF arguments give a formal structure in which informal statements (in natural language) are placed. A formal ETF argument uses the same structure to contain formal statements, i.e. statements in a formal language.

In SAM, the capability to support formal ETF arguments is provided by allowing ETF arguments containing informal or formal statements to be developed in separate but associated windows. To see the formal or informal version of a network of micro-arguments in a solution, the user may open a window for each.

Formal statements are expressed in a typed first-order logic language; the individuals to which they can refer and the types they can use are restricted to the content and metadata of the visible model views (i.e. the model views associated with the goal for which the formal argument provides a solution).

Entry and editing of formal statements is done using the same free-text editing facilities that are used for informal statements. Checking is done on request from the user, who can specify the kind (syntax, type or consistency checking) and scope (selected statements, the current window, all formal content associated with the current goal, all formal content in the current goal and all its subgoals, or all formal content in the safety case). This allows a user to ensure formal statements in arguments are syntactically correct before all the model content to which they refer is present.

4.2.7 Fault trees

Fault trees are a diagrammatic representation of the causal relationships between failure events in a system in terms of their effect on a particular event—the so-called "top event".

It is possible to interpret fault trees in ETF terms as a variety of warrant-establishing argument (providing the probability formula which would act as the warrant in an argument which calculates the probability of the top event) or as an argument which delivers a single claim (the statement of the probability of the top event). In SAM 3.2, we have chosen to take the latter view. We allow the creation and editing of fault trees and the user may then derive the minimal cutset formula for a tree, derive the corresponding probability formula, attach probabilities to events, and derive the probability of the top event.

4.2.8 Change management and hyperlinks

SAM keeps track of all creation and modification of the contents of the safety case. The life history of each part of the safety case is called its *provenance*, and is kept as a list of session records. Each session record contains the user name, the date and time of logging in, and the declared purpose of the session.

Any change to a solution or model view component may *jeopardise* components which depend on the changed one. If the user enables this feature, a jeopardised component will be indicated by a large triangle centred in the component's box. A small triangle will also appear in the upper right corner of the corresponding goal node. The jeopardy symbol indicates that the correctness of the statement in the component needs to be reviewed, because changes have occurred in other components on which this component depends. If the user makes any changes to a jeopardised component, further jeopardy will appear on any "downstream" components. In this way the system can systematically propagate the impact of changes up to the point where no further consequent change is necessary.

A change to an item in a window is also regarded as a change to the window and to the relevant goal, and a change in any goal is also regarded as a change to the goal lattice (i.e. to the goal lattice window).

Hyperlinks allow cross-references between one window's content and another's. This facilitates navigation between windows and allows jeopardy to propagate appropriately from one window to another.

4.2.9 Annotations

An annotation is a short note that is attached to a window. It allows the user to make a comment on the work contained in the window. Annotations are headed with the name of the user and the date and time. The user may only edit his or her own annotations, but may view all of them. He or she can specify whose annotations are to be visible; any goal whose content contains a visible annotation is marked with a small icon.

4.3 Design and Implementation Notes

4.3.1 Human-Computer Interaction

The SAM prototypes were developed on the Apple Macintosh, and followed the Apple Human Interface Guidelines. We strove throughout the project to ensure that the prototypes would be as usable as possible. It seems self-evident to us that if argumentation support is to be demonstrated as being viable, it must *inter alia* be demonstrated to be usable. Where we introduced new or rarely-seen interaction techniques (such as dragging from an ETF component to create another of the appropriate type) we sought user feedback on its ease of use and endeavoured to ensure consistency in all other parts of the prototype where the user might expect to find the same technique available (e.g. the goal structure editor).

We worked hard to develop a coherent conceptual model for the user interaction early in the project and to maintain this coherence as the prototypes evolved. Great care was taken over supporting this model in the interface through careful design of the menu structure and of the visible effects of commands. Some concepts—such as the idea of a single document containing the safety case—have been harder to support than others.

4.3.2 Internal Architecture

At the start of the project, alternative development environments were evaluated in order to find the best compromise between power (particularly in terms of access to all the Macintosh capabilities) and productivity. The chosen environment (Procyon Common Lisp) was strong on productivity, having a good implementation of Common Lisp and CLOS (the highest-level object-oriented language available), with incremental compilation and good debugging facilities. It was also quite good on trueness to the Macintosh. The main shortcoming was a lack of ready-made high-level application architecture, such as that provided by Apple's MacApp object-oriented application framework.

The tool-building end of the project (the project had parallel strands covering concept development, tool building and tool use/evaluation) began by developing a simple application framework, tailored to graphical interactive applications, containing classes such as Application, Document, Window and View, and supporting the Model-View-Controller style of user interface architecture. This gave the basic support needed for working with multiple views of multiple documents and carrying out graphical editing on their contents. A simple Toulmin form editor was demonstrated early in the project, and successive prototypes continued to make use of essentially the same object-oriented framework.

The early emphasis on graphical presentation of content has, however, had its drawbacks—the support for on-line graphical interaction has continued to dominate the internal organisation of SAM, and the support for good quality linear printed copy has suffered accordingly. Providing uniformly good support for a variety of concrete representations is difficult; one possible avenue for future investigation is to export a tagged textual form, such as SGML, which can then be handled by tools specialising in high-quality printed output.

4.3.3 *Alternative Back Ends*

For demonstration purposes, with small safety cases, it was desirable to use Macintosh files to hold complete safety cases. To meet concerns about scaling up the SAM technology for use on large safety cases, and to enable team support to be provided in the future, some kind of database server was clearly desirable. An evaluation of object-oriented databases was carried out, and ONTOS was chosen. After some exploration of ONTOS' capabilities, work began on defining a common interface to allow a given version of SAM to work with either files or databases. This turned out to be possible without significant disruption to the front end.

This common interface is now part of SAM, and the ONTOS back end has been proven in small-scale experiments, but problems with the RS/6000 implementation of ONTOS have prevented its use on a large scale.

5   Conclusions and Future Research Directions

The production of safety cases is both very important and very difficult. Without them it would be impossible to justify the deployment of many types of system, e.g. air traffic control systems, nuclear reactor protection systems. Many 'real-world' safety cases are tens of thousands of pages long, producing major issues of scale, as well as complexity. It is all too easy for inconsistencies to creep through in such a large volume of documentation.

The safety analysis tools currently available typically deal with specific techniques, such as fault-tree analysis. SAM is intended to build on the strengths of such tools, by providing a co-ordinating and integrating framework for complete safety cases. We know of no other work addressing these objectives.

The work in the ASAM project has resulted in an improved understanding of the overall structure of safety cases, the roles of the elements in this structure, and the use of semi-formal, as well as formal, representations of detailed chains of reasoning. The SAM tool support is tailored to this understanding of safety case structure. Several iterations of development (of ideas and of the tool itself) have taken place, with evaluation against realistic (but small) examples at each stage, and the results of evaluation have been used to influence the ideas and the tool design. The ASAM project has been largely successful, but there is still some way to go before SAM's concepts and technology can be put into use on real safety cases, on a large scale.

So far, SAM makes few assumptions about (and so gives little support to) the *process* of safety case development. The notations used, and SAM itself, have only been tried out on small samples of safety analysis, in scenarios involving single users, working on static snapshots of design. Real safety cases are large, represent the work of a team of people, and may have to respond to changes in a design which is being developed in parallel with, and to some extent as a result of, the safety analysis. To provide computer-based support for work on this scale further work is required to understand, model, and design support for the interleaved processes of safety analysis and design. The resulting tool support will have to be evaluated using large enough case studies, and large enough groups of users, to provide a real test of the process support facilities. Further work is also needed on linearising the safety case content to produce a good printed version which still corresponds clearly to the on-line version. This is currently the biggest impediment to small-scale use of SAM. Finally, the impact on safety of the use of SAM in safety analysis will need to be analysed—SAM, the process it supports, or both may need refinement as a result.

References

[HSE 87]   Programmable Electronic Systems in Safety Related
                Applications, vol. 2, General Technical Guidelines,
                Health and Safety Executive, HMSO, London, 1987.

[Jones-Lee 87] The Political Economy of Physical Risk,
                M.W. Jones-Lee, J. Soc. Radiol. Prot., 7 (1) (1987).

[Leveson 86]   Software Safety: What, Why and How,
                ACM Computing Surveys, 1986.

[McDermid 92]   Safety Cases and Safety Arguments, J.A. McDermid,
                in Proceedings of CSR Conference on Software Safety,
                Luxembourg, April 1992.

[Storrs 91]   The Policy System, G. Storrs,
                In Knowledge-Based Systems and Legal Applications,
                T. Bench-Capon (ed), Academic Press, 1991.

[Toulmin 58]   The Uses of Argument, S.E. Toulmin,
                Cambridge University Press, 1958.

[Toulmin 84]   An Introduction to Reasoning,
                S.E. Toulmin, R. Rieke and A. Janik,
                Macmillan, New York, 1984.

The Need for Evidence from Disparate Sources to Evaluate Software Safety

Bev Littlewood

Centre for Software Reliability, City University
London, UK

Abstract

A system may fail because of its engineering hardware, computer hardware, computer software, or because of a human component. The impact on overall system dependability of hardware is well understood, provided that it is free from design faults. Furthermore, we can often engineer our systems so that the impact of these sources of unreliability is negligible.

However, some hardware failures and *all* software failures are due to design faults. Reliability in the presence of design faults and human operator errors is not well understood. This poses acute problems for the assessment of safety-critical systems in the presence of the effects of human errors, made during the design process or during operation. It can easily be shown that only modest reliability can be demonstrated by the direct observation of the system in test or operation. In this paper we discuss these problems in detail and consider some ways in which evidence from other sources might be used to increase our confidence. This work forms the basis of the project *DATUM: Dependability Assessment of Safety-critical Systems Through the Unification of Measurable Evidence.*[1]

1   Introduction

It is our contention that safety assessments should ideally be *quantitative*. For a fly-by-wire airliner, we need to know that the flight control system has an acceptably low failure rate; for an air traffic control system there might be a similar requirement with, in addition, a need for high availability; for a reactor protection system, the probability of failure on demand must be sufficiently low. We believe it is inevitable that these quantitative requirements for the safety of such systems be expressed in *probabilistic* terms, because of the inherent uncertainty about the behaviour of all real systems during operational use.

It is generally agreed that the major problem facing anyone with responsibility for deciding whether a safety-critical system is fit for purpose - safe *enough* - is in quantifying the contribution to its failure behaviour of human errors, manifested both as design faults and as operator faults. For the most part, work in this area has not concentrated upon the problem of safety *per se*, but rather has seen safety as merely the most difficult end of a spectrum of general problems related to evaluation. Most of this work has concentrated upon software, probably because this represents the problem of design faults in its purest form, since software faults are *always* design faults. However, safety is quintessentially a *system* problem, and it seems inevitable that our judgements about dependability will need to take account of many different sources and types of

---

[1]   The *DATUM* project involves Centre for Software Reliability and Centre for Human Computer Interface Design (both at City University), Lloyds' Register, and Royal Holloway and Bedford New College. It is supported by SERC as part of the DTI/SERC Safety Critical Systems Programme. The author thanks his colleagues for their valuable input to this paper: Norman Fenton, Victoria Stavridou, Alistair Sutcliffe, Roger Shaw, Martin Neill.

evidence, shown schematically in Figure 1. The sources of information may be:

- the system viewed as a *product*, such as its failure behaviour under specific test conditions
- the *process* by which the system was developed, such as the use of formal methods or fault tolerant design
- the *resources* used to create the system, such as the experience of key development staff, or the quality of a compiler
- the way *humans* interact with the system or as part of the system, such as design features of the HCI

The types of information are generally

- scientifically rigorous quantification, involving measurement
- informed engineering judgement, perhaps including expert, subjective opinion.

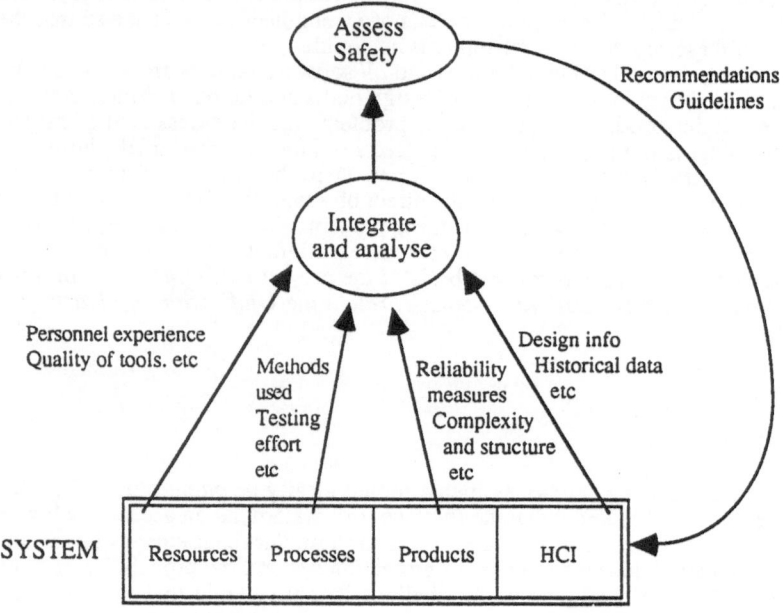

Figure 1: Disparate sources of evidence about the reliability and safety of a system, and their analysis and composition for overall system assessment.

Unfortunately, current practice is overly dependent upon the latter and this is often conducted in a very informal manner. This is exemplified in recently proposed standards relating to the development of safety-critical systems. These are primarily process and resource based, implying that having the right people and using the right methods will enable one to have confidence in the dependability of a system. Unfortunately, there is almost no empirical evidence to confirm that specific recommended techniques, such as formal methods or fault tolerance, can ensure an adequate level of safety. There is more evidence about the likely impact on safety of specific factors in human computer interaction. In particular, research has produced a wealth of ergonomic guidelines which aim to improve interface designs, as well as evaluation methods for detection of usability problems.

The most relevant and highly developed techniques in this area are those for estimating and predicting software reliability based on the observation of the failure process during testing and operation. Provided that failures *are* observed, it is possible to obtain accurate reliability measures and know that confidence in such figures is justified. However, some safety-critical systems have extremely high reliability requirements and in such circumstances any reliability estimates obtained merely from direct observation of the failure process are likely to fall far short of what is needed.

There has also been some quantitative work which has attempted to relate measures of internal system properties to operational reliability. However, there has been almost no work attempting to combine these and other sources and types of information into quantitative assessments of overall system safety.

The work in *DATUM* will be in two main areas. Firstly there are problems associated with the extraction of quantitative evidence from different sources, secondly there is the problem of integrating this disparate evidence into quantitative assessments of system safety and reliability. There is not space in this paper to describe the whole *DATUM* project; instead we shall select some of the more important and innovative features.

2   Background: the problem

A great deal of work has been carried out on the problem of estimating and predicting the reliability of a program as it is being debugged during test: the *reliability growth* problem. There are now many stochastic models which purport to be able to provide such predictions [Goel and Okumoto 1979; Jelinski and Moranda 1972; Littlewood 1981; Littlewood and Verrall 1973; Musa 1975; Musa and Okumoto 1984]; see [Xie 1991] for a useful survey of this work. Although none of these can be relied upon always to give reliability measures that are accurate, recent techniques allow us to check whether they are providing accurate results on a particular software system [Abdel-Ghaly, Chan et al. 1986; Littlewood 1988]. It is often possible to use these techniques to allow a model to 'learn' from its past errors and so recalibrate future reliability predictions [Brocklehurst, Chan et al. 1990]. The bottom line is that it is usually possible to obtain accurate reliability measures from software reliability growth data, and to know that confidence in such figures is justified.

Another area where modelling has had some success is in the incorporation of structural information about the program into the reliability estimation process. These models aim to emulate the classical hardware procedures which allow the reliability of a system to be computed from knowledge of the reliabilities of its constituent components, together with information about the organising structure [Barlow and Proschan 1975]. Whilst these hardware theories are essentially static, the software approach has to be dynamic and emulate the way in which software components (modules) are successively exercised in time. This is done by assuming Markovian [Littlewood 1975; Siegrist 1988a; Siegrist 1988b] or semi-Markovian [Littlewood 1976; Littlewood 1979] exchanges of control between modules. Each module can itself fail with its own unique failure rate, and the exchanges of control between modules are failure-prone in the most general formulation. The potential advantage of this approach is that it allows the reliability of a system to be predicted before it is built, in the event that it is to be built of modules whose failure history in previous use is known.

Other structural models have been studied in order to model the failure behaviour of fault-tolerant software based on the idea of design diversity. The great difficulty here is that we know from experimental studies [Eckhardt, Caglayan et al. 1991; Knight and Leveson 1986a] that it would be too optimistic to assume that diverse software versions fail independently. This precludes the simple modelling assumptions that are sometimes made in the case of hardware *redundancy*. Estimating the actual degree of dependence in failure behaviour between two 'diverse' software versions seems very difficult; indeed, it seems as hard as simply treating the whole fault-tolerant system as a black box and estimating its reliability directly by observing its failure behaviour [Miller 1989].

These are some of the areas where there have been advances recently in our ability to measure and predict the reliability of software. The important point from a safety-critical perspective, however, is that this success story relates only to those cases where the reliability being measured is quite modest. It is easy to demonstrate that reliability growth techniques are not plausible ways of acquiring confidence that a program is ultra-reliable [Littlewood 1991; Littlewood and Strigini 1991; Parnas, Schowan et al. 1990]: the testing times needed become astronomically large as a result of a law of diminishing returns, and the issue of whether the test inputs are truly representative of those the system will meet in operational use becomes serious. Similarly, as we have seen, the effectiveness of fault tolerance is limited by the degree of dependence in the failure processes of the different versions, and experiments suggest that this will be significant.

If we really need an assurance of ultra-high *system* reliability, and this appears to be inescapable in some safety-critical applications, it seems that this will have to be achieved without depending upon *software* to be ultra-reliable. In fact, of course, the problem is even worse, since everything we have said about software applies to design faults in general. Any claims that particular systems are safe because their control systems are ultra-reliable must take note of these unpalatable facts. The only possible exceptions are those systems that are so simple that it can be argued that they are completely correct (and are a complete and accurate embodiment of their high level requirements, which must also be extremely simple). This observation may give us some leeway to build computer-based systems, with the extra non-safety functionality that these can deliver, that are nevertheless measurably safe, by confining the safety-critical functionality in a tightly controlled kernel. The subsequent evaluation of such systems seems a potentially fruitful area for further work in *DATUM*.

Otherwise, the major theme of the project will be an attempt to push back the present limitations of measurement by incorporating information from other sources than the ones that have been outlined above. Any well managed project to develop a safety-critical system will have available a wealth of such potentially important information which could be used. This information may itself be quantitative, such as: fault and change reports from various project phases; test results including coverage measures; outputs from static analysis or metrics tools; various internal system quality indicators such as modularity measures; measures of effort associated with various project phases; historical evidence of efficacy of tools and techniques. [Fenton 1991] provides a comprehensive overview of such software measures. Alternatively, the information may be qualitative, such as: minutes of formal review meetings; verification arguments; information about relevant experience of key personnel; general information about quality of tools and methods used. We are especially concerned with information from the system development process and from an evaluation of the effects of humans within the system.

The use of formal methods is widely advocated as a way of increasing confidence in the dependability of safety-critical systems. Although there is a clear prima facie case for consenting to this view there is, in general, a dearth of facts to support it. For example, it is often stated that testing coupled with diversity can only yield a failure rate of about 10^{-5} per hour, and thus the use of formal methods is advocated for reaching the required ultra-high reliability levels of over 10^{-9} [Moser and Melliar-Smith 1990]. There is no hard experimental evidence to support this view. This situation is most unsatisfactory since the key notion in dependability is that *reliance must be justifiable*. This means that we need explicit and testable requirements as well as credible analytical and experimental evidence demonstrating the satisfaction of the requirements by the system. Such evidence is simply not available for applications of formal methods technology.

Whilst some comparative experimental evidence on the effectiveness of techniques such as testing and fault tolerance is available [FAA 1982; Knight and Leveson 1986b], the assessment, measurement and prediction of the contribution of formal methods to safety is a subject largely unresearched. There are two major reasons for this. First, there is very little experimental data. This is because there are relatively few instances of use of formal methods for safety-critical systems and even in these cases there is no systematic data collection [Bowen and Stavridou 1992]. But perhaps more fundamentally, quantitative evidence is sometimes perceived as inappropriate for qualitative improvement techniques such as formal methods. The two major schools of thought concerning risk analysis and assessment, qualitative and quantitative, are often at odds with each other. The quantitative school believes that probabilities are primarily reflections of the actual frequency of events, thus are objective and can be used for predictions of future events.

The qualitative school does not play the 'numbers game': rather, they believe, that risk is minimised and contained not by statistical test programs, but by paying particular attention to the details involved within a system's design. In fact, this represents a confusion between the goal of *achieving* dependability, and that of *measuring* what has been achieved. It may very well be true that formal methods are a good means of achievement, but they leave open the question of exactly what has been achieved in a particular case. Equally importantly, any claims for efficacy can only be substantiated scientifically if they can be shown quantitatively to deliver that which they promise. In fact this debate need not be so confrontational and sterile as it has been. It is not necessary for the qualitative school to eschew those beliefs about the efficacy of formality that are based upon a trust in the logical and mathematical: these beliefs should themselves be quantifiable and the results open to testing against the observable facts.

In common with most other formal methods research workers (e.g.[Cohn 1988]), we believe that while formal specification and proof successfully address some aspects of dependability, they are not a panacea. Furthermore, proof and measurement are not at odds; they are in fact complementary and one does not make sense in the absence of the other. This necessity is quite obvious when one reflects on the extremely low failure rates required by some applications [FAA 1982]; the challenge is to be able to gain confidence that such goals have been reached via evidence from proof, measurement or any other source.

One of the main failings of dependability assessment has been the absence of integrated system models which consider the role of hardware, software and people as determinants of overall system reliability. At present, for example, the allocation of both the degree and nature of the responsibility between software, hardware and humans is conducted quite informally, and occasionally results in quite startling differences between different systems. As an example, the Airbus A320 design places great trust in the flight-control computer system, to the extent of allowing only rare pilot over-ride, and the simple hardware back-up is merely vestigial; whereas the Sizewell B computer-based primary protection system is backed-up by a purely hardware secondary system with extensive (although not complete) functionality, and a major aim of the design has been stated to be 'giving the operator time to think'. Although the former is a continuously functioning *control* system, whilst the latter is a *safety* system that is expected to be called upon only rarely, these differences in rôle cannot account entirely for such radically different design philosophies, and in particular the apparently different degrees of trust placed in both people and software. We need ways in which designers can adjudicate between such different approaches, taking into account both the specific nature of the application and the plausible degrees of trust that can be allocated to hardware, software and humans. Many studies have been concerned with diagnosis of human failure motivated *post hoc* as a result of accidents. These studies have demonstrated that reasons for failure are complex and idiosyncratic. Unfortunately, such analyses are limited in their predictive power, whereas designers need to anticipate the reasons for human failure and incorporate design features to minimise the potential for human error.

Research into the rôles played by humans has produced a wealth of ergonomic guidelines which aim to improve interface designs and evaluation methods for detection of usability problems [Galer 1987; Whiteside, Bennett et al. 1988]. Currently, HCI can offer advice which should prevent errors in interaction and methods for assessment of design problems. Guidelines and usability evaluations both require human factors experts for interpretation, however, quantified evidence for the dependability of such expert audits is rare. Furthermore, given that the supply of human factors experts is limited, assessment techniques should be devised for use by less highly qualified staff. Some progress has been made by checklist evaluations (e.g. [Ravden and Johnson 1989]) which provide lists of design principles and guidelines which should be followed to prevent errors. However, interpreting checklists in practice is problematic [Sutcliffe and Springett 1992]. An alternative approach is to use design walkthrough techniques borrowed from software engineering [Lewis, Polson et al. 1990], but these techniques can only pinpoint problematic features of specific designs and do not provide more general predictions of dependability.

Another approach is to describe general situations which can be recognised as having the potential for certain types of error in human computer interaction. Generic models have been proposed in requirements engineering as clichés [Reubenstein 1990], and abstract domain models [Maiden and Sutcliffe 1992]. These models have been used in process control domains by [Sutcliffe and N.A.M. 1991] (see also [Woods 1988]) who describe properties of typical tasks (e.g. monitoring, decision making, scheduling). Generic models could form the input data for evaluating HCI errors by cognitive walkthroughs [Lewis, Polson et al. 1990] although such techniques currently lack principled models of human information processing to enable prediction of failure.

Understanding human error requires cognitive models for interpretation of interaction. Relevant models exist for failure in human memory [Reason 1990], attention [Norman and Shallice 1986], problem solving and task learning [Anderson 1985; Rassmussen 1986]; however, these models describe general cognitive mechanisms and do not have sufficient predictive power to pinpoint potential errors in specific contexts. More targeted models of human reasoning and operation have been developed by researchers in the dynamic tasks area, ranging from taxonomies of behaviours and decision making [Rouse 1981], to descriptive models of reasoning and behaviour [Woods 1988]. Although these models can give some indication about the sources of error they do not explicitly consider how human failure occurs. Variants of this approach with more predictive power have modelled human information processing and its known limitations in terms of working memory, attention, and cognitive resources [Bainbridge 1992; Grant 1992]. These models are developing towards a concept of context, i.e. identifiable situations in which human behaviour may be predicted for specific types of task.

The cognitive science and ergonomics literature contains a wealth of detail about human abilities within narrow, experimentally proscribed, contexts. Qualities of human perception and motor control are well understood and can be generalised into predictions about low level attentional and perceptual behaviour [Norman and Shallice 1986; Wickens 1984]. It is more difficult to make predictions about human decision making, although the critical influences on human reasoning are known from HCI usability studies [Lewis, Polson et al. 1990] and predictions of failure could be derived from cognitive models of interaction (e.g. Interacting Cognitive Sub Systems [Barnard 1987], Programmable User Models [Young, Green et al. 1989]). Cognitive science, therefore, has reached a state of maturity where existing models may be synthesised and applied to the prediction of human error at the level of general contexts. While this will not enable exhaustive prediction of human error, it can at least indicate the type and likelihood of failure in certain situations.

A major theme of *DATUM* will be to bring together the disparate sources of evidence about a safety-critical system; the aim being to provide a means whereby we can arrive at levels of confidence that are greater than those that we could obtain from any single one of the sources of evidence that have already been discussed. There is little existing formal work on this problem, at least in the context of dependability measurement and prediction. However, we know that those responsible for building and assessing safety-critical systems do this all the time: it is called 'engineering judgement'. However, there is evidence that human judgement, even from expert subjects, shows fairly consistent bias when unaided by a formal framework that can check for such errors [Henrion and Fischhoff 1986]. It is interesting that these judgements err in two ways: experts tend to give a view of their likely achievement that is too *optimistic*, and they *underestimate* their own chances of error in the making of such judgements. We see no reason to believe that 'engineering judgement' about the dependability of software (or any complex design) will not suffer from these errors. However, there do exist theories which permit a rigorous treatment of human beliefs in the face of uncertainty: Bayesian probability [Aitchison and Dunsmore 1975; DeGroot 1970], possibility theory [Dubois and Prade 1988] and Shafer-Dempster theory [Shafer 1976] are examples.

Ultimately, human judgement in the design process is influenced by many factors, such as individual ability, techniques employed and problem complexity. While it is not currently tractable to quantify the reliability of designers, useful assessment can be made about the procedures they follow and of the likelihood of their adherence to these procedures. For example, human reaction to the introduction of structured methods has been a proliferation of individual tailoring of methods. If formal methods assert a certain reliability, what is the reliability of human conformance? There is a dearth of data on reliability impact of formality in the design process. Again theories of human belief can be employed to model judgements made during design, as well as assessment of existing products.

3   Some research themes in *DATUM*

In this section we shall give a brief account of some of the specific ways in which these problems - of judgement, of composition of evidence, in the presence of human errors both in operation and in design - will be addressed in the *DATUM* project.

3.1   *Judgement and the composition of disparate evidence.*

Realism suggests that future system builders are not going to restrict themselves to building safety-critical systems which can be demonstrated by direct measurement to have achieved their required safety: we know that this can only be done for relatively modest levels [Littlewood 1991; Littlewood and Strigini 1991], whereas systems are already being built with requirements that are several orders of magnitude beyond what is measurable in this way. On the other hand, it is not necessary to relinquish scientific rigour even if we wish to incorporate into our judgements of systems an element of subjectivity. Between the two poles of complete subjectivity, or 'pure' engineering judgement, on the one hand, and direct measurement on the other, lies an area that has not so far received much attention in system modelling: the rigorous composition of evidence from disparate sources into quantitative (probabilistic) measures of system safety and reliability.

Since one of the sources of such evidence *is* human judgement, we shall start our investigation with an examination of the accuracy of this in those fields where it might be expected that there is extensive experience. We shall try to determine whether human judges in the area of safety-critical systems can be assumed to be well-calibrated, and if not whether there might be opportunities for recalibrating them in a manner akin to that developed for the software reliability growth models [Brocklehurst, Chan et al. 1990]. Notice that the issue of judgement is one that affects not only individual judges, but also *communities* as represented, for example, in the claim limits that are embedded in standards [MoD 1989].

Formal quantitative evaluation of judgement requires means of elicitation that will provide numerical results in forms that will facilitate the composition of this evidence with that from other sources. We believe that Bayesian probability might be an appropriate mechanism here, but will also investigate other avenues such as possibility theory, Shafer-Dempster theory.

When we come to consider the problem of combining the different kinds of evidence that are available about the safety of a particular system, it seems clear that there will be important interactions and dependencies that we ignore at our peril. For example, little is known about the way in which the degree of dependence of failure behaviour between diverse versions in a fault tolerant architecture will change as the different versions are tested and debugged. Do the kinds of faults being experienced by the different versions become more similar or less? Issues of this kind are clearly vitally important and must be resolved before we could combine evidence from testing with evidence about the nature of the fault-tolerant architecture. It may be that the best we can do here is establish believably conservative claim limits, but it is important that these are at least expressed quantitatively. Similar dependency issues arise in other areas. For example, having a human operator able to intervene in system operation is a kind of diversity: are there similar issues of dependence here as in more conventional design diversity? Can we claim that operator errors are different in kind from those made by designers, and if so, to what degree?

The specific modes of composition that are appropriate will depend upon answers to questions like these. Once again, this composition may be possible in the Bayesian framework by regarding the different kinds of evidence as being presented sequentially to the person (or possibly computer program) who has the responsibility to form the final judgement about the safety of the system under examination. Thus initial prior beliefs will be transformed by one set of evidence, becoming the prior belief for the next set of evidence, etc. This Bayesian approach has the advantage of conceptual simplicity, but it may not be ideal for all applications. It tends to 'average out' uncertainty, for example, and the Shafer-Dempster approach may sometimes be preferable since this retains both 'belief' and 'strength of belief'. Comparison of the several different conceptual approaches here will be an important part of the study. It seems likely that no particular one is universally 'best', but that choice will to a large extent be pragmatic and dependent upon the end use of the safety measures. In the case of certification, for example, there may exist legal requirements that will force a particular approach. Our purpose will be to delineate the relative merits and demerits of the different candidates.

If we can arrive at answers to some of these questions of compositionality, we shall be able to have more scientifically plausible measures of the safety of existing systems. We shall also be able to provide a more formal framework for decision-making at the early stages of system design, to help answer, for example, those crucial questions about the trade-off between hardware, software and humans.

3.2 Assessment techniques for reliability in the presence of human errors.

One of the most successful areas of research in software reliability has been in reliability growth modelling: estimating and predicting the growth in reliability that arises as a result of fault removal. In fact reliability growth is not only observed in an individual system for this reason, but also as a result of operator errors becoming less frequent with experience, faults in operating procedures being discovered and removed, etc.

There is also evidence of reliability growth between successive *generations* of equipment, not only in electronic systems but also in heavy engineering components such as steam generators and in aircraft. Data from the CAA show a steady growth in the reliability of aircraft with a factor of 5 improvement in the fatal accident rate in jet aircraft between 1960 and 1980. Although this growth in safety must have been accompanied by an increasing sophistication and complexity in the product concerned, there are other examples where the rate of innovation was too fast and new products were less reliable than the old ones.

Common sense would lead us to suspect that products in which there is much innovation between generations would be prone to design faults. This is indeed the case with computer control equipment, where the size of the software systems can increase by an order of magnitude between product generations, but the example of aircraft development suggests that complexity alone may not necessarily incur the penalties of lower reliability. Here there is operating some kind of learning process from one generation to another: if nothing else, a simple increase in general scientific and engineering understanding as time passes. We shall attempt to investigate the trade-offs that are taking place here between innovation and the unknown on the one hand, and learning on the other. A useful starting point might be an investigation of the accuracy of the predictions of the system developers themselves.

One way in which we can carry over confidence into a new product from past experience with earlier ones is via reuse of intellectual property. Thus a software module that has been in use in earlier systems will have built up a history of use that should allow its contribution to the unreliability of a new system to be computed. There already exist probability models that allow the system reliability to be computed from the reliabilities of the component modules [Littlewood 1979]. However, there has been little operational vindication of this approach, and there must be some concern that the assumptions are a little unrealistic: for example, it is not clear that the reliability of a module in a novel application (i.e. in a new architecture) will be the same as it was in the environment in which the failure data used in the estimation of its reliability was collected. This is an area where further research is needed: in the event that this assumption is violated, it may still be possible to use past evidence to estimate the reliability of a module in a novel application so long as we have sufficient information about the differing natures of the present environment and the ones from which failure data were collected.

Fault tolerance in software introduces some other novel modelling problems for real-time process control situations that may yield to investigation. For example, many continuous time control problems often work on a discrete time basis determined by the cycle times of sensors and scanners. A particular version will typically execute many of these cycles successfully before failing when it meets an appropriate conjunction of sensor values on a particular cycle. For most processes that are controlled by such a system, it is likely that the immediately succeeding cycles to the failed one will receive sensor input that is 'close' to that received during the failed cycle, and these are also likely to fail. Thus we might expect to see failures occur not singly, but in clusters. Each version in a fault tolerant architecture will exhibit such a clustered failure process. When these complex failure processes are brought together in the fault-tolerant scheme, via some adjudication mechanism such as a voter, the output will also tend to be clustered, but in some different way from the individual processes. The nature of this output failure process is, of course, of vital interest: the physical integrity of the controlled plant may allow individual failed cycles, or even small failure clusters, without catastrophic results, but will not tolerate large ones. Work has begun on these problems [Csenki 1989a; Csenki 1989b; Csenki 1989c], but there are still important gaps that need addressing.

Recent work on modelling of the dependence of the different version failure processes in a fault-tolerant architecture has introduced the interesting idea that, although it is impossible to achieve independence via diverse development, it may be possible to *force* diversity and get even better results than would have been obtained with independence [Littlewood and Miller 1989]. So far, this is merely a qualitative theory, but if it were possible to quantify the benefits in a particular application, there would be considerable improvement in the dependability levels that could be assured.

3.3 Development of system level models of dependability

The objective here is to investigate what factors should be measured in human-computer systems and how these factors inter-relate. The system will be decomposed into a model of human dependability and a further model of human computer interaction. The approach taken is to develop a bridging model of human information processing and interaction models from the existing literature. The concept of bridging models is to sacrifice some cognitive validity of basic memory and problem solving models (cf. [Anderson 1985]) to enable more predictability by specialisation with domain specific features. The model will be based on components which describe cognitive resources such as working memory [Hitch 1987], contention in perceptual channels [Wickens 1984], constraints on processing, i.e. attention [Norman and Shallice 1986], and information processing resource limitations (from Barnard's 1987 ICS model). The resulting model will describe human information processing and pinpoint key aspects which can cause failure within the human sub system. e.g.

- attentional failure in perception or cognitive action
- failure caused by contention between resources in perception
- problems caused by limitations on resources such as working memory
- possible influences on failure of reasoning from memory, experience and the environment.

The model will be used in conjunction with input data consisting of general task contexts. Work by Bainbridge [Bainbridge 1992] and Grant [Grant 1992] has suggested that in automated and semi learned tasks, people structure activity in a sequence of contexts. This follows the tradition of skill based learning [Anderson 1985; Rassmussen 1986] which proposes that skilled behaviour is run as automated procedures triggered by an environmental situation. Theories of errors have accounted for a range of human failure which occur in context mismatching and incorrect triggering [Norman 1988; Reason 1990]. It is reasonable to expect, therefore, that a set of general contexts and error risks could be described. Properties of command and control tasks could be described with these contexts and assessed for dependability of human performance. We propose to define a set of such contexts (cf. scenarios [Young and Barnard 1987]) to describe key features of situations which have to be evaluated. These descriptions will become input data for the model of human information processing to make predictions about potential failure. The level of prediction will be prescriptive enough for dependability assessment of high level design features. Mapping generic models to specific applications should facilitate assessment of the dependability of design features in terms of their usability and potential for failure.

3.4 Assessing judgement of dependability.

The ultimate judgement as to whether or not a particular system is sufficiently safe that it can be deployed currently often takes place in a framework which is formal, but not overtly quantitative. In fact, of course, there is always a quantitative requirement for the safety of a system, even in those cases where this is not stated overtly. It seems likely that this kind of human judgement will continue to play a large rôle in the future, and we therefore propose to see how accurate such judgements actually are when compared with the actual numerical safety or reliability. Firstly we propose to look at historical data, comparing the judgements made of safety-critical systems prior to operational use with their later observed actual safety. Secondly, we propose to conduct experiments on human judges ourselves; here it will be necessary to work with systems with rather modest reliability requirements. In all these areas of judgement, we shall investigate the possibility of recalibrating the assessors by comparing the predictions with the outcomes, using techniques similar to those developed for the reliability growth models [Brocklehurst, Chan et al. 1990].

A particularly important area of concern here is judgement about human computer interfaces. Current techniques for auditing the quality and inter alia dependability of interfaces depend largely on expert judgement, although some observation and semi quantitative techniques can give more objective measures [Sutcliffe and Springett 1992]. We propose to assess the reliability judgement of human factors experts' judgement by taking usability assessment to approximate for dependability. HCI evaluative tools will be developed using off the shelf usability guidelines [Galer 1987; Sanders and McCormick 1987] to produce check list inventories of HCI design principles and guidelines which should be followed. This approach, following [Ravden and Johnson 1989], is a useful first step although interpretation of checklist evaluation results is problematic [Sutcliffe and Springett 1992]. We shall augment the checklist approach with usability walkthrough techniques to assess error potential at each step of interaction drawing on evaluation techniques from the York Manual [Wright and Monk 1989] and work on contextual evaluation [Lewis, Polson et al. 1990; Whiteside, Bennett et al. 1988]. This will provide a battery of assessment methods which may be applied. Use of these methods by human factors experts in case studies will provide predictions of dependability. These predictions will then be validated against historical data to assess the reliability of expert judgement. Different levels of expert judgement will be investigated i.e. consensus among a cohort of human factors experts using one technique, judgement by experts using different techniques. Resources and subject availability will preclude controlled experiments so more economic techniques such as scenarios, structured interviews and protocol analysis will be used.

3.5 Modelling the designer

Here we shall investigate human dependability in design. While detailed assessment is impossible given the creative and variable nature of design, we believe that useful assessment criteria can be researched. In particular conformance of designers to design techniques should have estimable consequences in terms of product reliability. Our focus will be the application of formal methods. This task will analyse the application of formal methods by first developing a framework of application and then collecting data from designers practice.

We make a key distinction between two main objectives in formal design:

- Validation - establishing the accuracy and completeness of the system requirements with the users. This is a process of negotiating and iterative understanding with may be helped by formalism.

- Verification - correctness proof of software behaviour, the typical usage of formal methods in proving invariants, consistency, etc.

The framework will address different aspects of the design process, e.g. validation of requirements with users, verification of specifications and designs, reliability of transformation applied in the design process, etc. Considering no canonical view of the design process exists, we will take the procedure of a typical formal method (e.g. VDM, Z), as a working example. The conformance of designers with the method will be tested by assessing performance in key tasks:

- representation of an application problem in a formal notation
- validation of the notation to check user requirements
- verification of the formal specification with theorem provers
- transformation of specifications during design

As with assessment of HCI, expert resources will preclude a controlled experiment so more economical empirical techniques will be used. Observer bias is well known in methodology studies so non-intrusive techniques will be employed in an attempt to capture realistic data on the effectiveness of formal method in industry and academic settings.

3.6 Models of the design process

The objective of this task is to develop models of dependability assessment for alternative design paradigms. In particular reuse, and module evolution through versions, seem plausible candidates for achieving reliability. They may even allow more accurate *evaluation* of reliability by taking account of previous experience of use. Another concern is reliability assessment in fault tolerant architectures. Reliability of software architectures can only be estimated in terms of what is known about the systems. We will investigate two aspects:

- Reliability of interfaces and of inter module communication in terms of formal specification
- Reliability of module designs in terms of formal specification

Inter module coupling is well known to have reliability implications. It should therefore be possible, given module interface specifications, to formally assess dependability of module configurations. The concept of generic system types can be used to describe template designs. These boilerplate configurations can be assessed for a baseline dependability, and then the effect of adding further modules could be evaluated in 'what if' predictions for dependability in systems designed with reusable modules. The same principle can be applied to software evolution, in which baseline dependability will exist from real data on the extant system. The hazard of introducing new versions can be assessed by formal verification of the changes imposed.

Fault tolerant architectures are amenable to the same treatment. Different configurations can be modelled in terms of their dependability properties when fully functional and under various stages of degradation. Assessment of complex systems will have degrees of tolerance. Formality as an assessment criterion can only be applied to elements of the system which are accessible and accurately documented. Hence a two layered approach may be taken to specify intra-module reliability criteria which are either present or absent (e.g. degrees of formal specification of the module internals and interface) and inter-module dependability in terms of formal assessment of interprocess communication.

4   Summary

The programme outlined here forms the major part of the research that is planned for the 3-year *DATUM* project, addressing the problem of extending our capacity to give quantitative assurance of the dependability of complex safety-critical systems containing both computers and humans. Other activities in the project include the development of prototype tools to support data collection and interpretation, and the development and evaluation of a prototype standard for the measurement-based assessment of software. We also expect to devote some effort to promotion of this measurement-based standard, and in particular to lobby for change in the way that system dependability is assessed in wider, non-critical applications, where we expect that much of this work will find applicability.

Readers who are interested in this research project are invited to contact the authors for further details. In particular, we would like to make contact with owners and builders of systems which might be suitable subjects for study during the period of the research programme.

References

[Abdel-Ghaly, Chan et al. 1986] A.A. Abdel-Ghaly, P.Y. Chan and B. Littlewood, "Evaluation of Competing Software Reliability Predictions," *IEEE Trans. on Software Engineering*, vol. SE-12, no. 9, pp.950-967, 1986.

[Aitchison and Dunsmore 1975] J. Aitchison and I.R. Dunsmore. *Statisitical Prediction Analysis*, Cambridge, Cambridge University Press, 1975.

[Anderson 1985] J.R. Anderson. *Cognitive Psychology and its Implications*, New York, W. H. Freeman, 1985.

[Bainbridge 1992] L. Bainbridge. "Mental models and industrial process operation," in *Models of the Mind*, pp. 119-143, Academic Press, 1992.

[Barlow and Proschan 1975] R.E. Barlow and F. Proschan. *Statistical Theory of Reliability and Life Testing*, New York, Holt, Rinehart and Winston, 1975, 290 p.

[Barnard 1987] P.J. Barnard. "Cognitive resources and the learning of human computer dialogues," in *Interfacing Thought: Cognitive Aspect of Human Computer Interaction*, MIT Press, 1987.

[Bowen and Stavridou 1992] J. Bowen and V. Stavridou. *Safety-critical systems, formal methods and standards*, PRG-TR-5-92, Programming Research Group, Oxford University Computing Laboratory, 1992.

[Brocklehurst, Chan et al. 1990] S. Brocklehurst, P.Y. Chan, B. Littlewood and J. Snell, "Recalibrating software reliability models," *IEEE Trans Software Engineering*, vol. SE-16. no. 4, pp.458-470, 1990.

[Cohn 1988] A.J. Cohn. "Correctness properties of the Viper block model: the second level," in *2nd Banff Workshop on Hardware Verification*, Springer Verlag, 1988.

[Csenki 1989a] A. Csenki. "Recovery block reliability analysis with failure clustering," in *Dependable Computing for Critical applications*, Santa Barbara, Ca, 1989a.

[Csenki 1989b] A. Csenki. *Recovery block reliability modelling with nested clusters of failure points*, Centre for Software Reliability, City University, London, 1989b.

[Csenki 1989c] A. Csenki. *Reliability models of fault tolerant software*, Centre for Software Reliability, City University, London, 1989c.

[DeGroot 1970] M.H. DeGroot. *Optimal Statistical Decisions*, New York, McGraw-Hill, 1970.

[Dubois and Prade 1988] D. Dubois and H. Prade. *Possibility Theory: An Approach to Computerised Processing of Uncertainty*, New York, Plenum Press, 1988.

[Eckhardt, Caglayan et al. 1991] D.E. Eckhardt, A.K. Caglayan, J.C. Knoght, D.F.M. L. D. Lee, M.A. Vouk and J.P.J. Kelly, "An experimental evaluation of software redundancy as a strategy for improving reliability," *IEEE Trans Software Eng*, vol. SE-17, no. 7, pp.692-702, 1991.

[FAA 1982] FAA. *System Design Analysis*, 25.1309-2, US Department of Transportation, Federal Aviation Administration, 1982.

[Fenton 1991] N.E. Fenton. *Software Metrics: A Rigorous Approach*, London, Chapman and Hall, 1991.

[Galer 1987] I. Galer. *Applied Ergonomics Handbook*, Butterworths, 1987.

[Goel and Okumoto 1979] A.L. Goel and K. Okumoto, "Time-Dependent Error-Detection Rate Model for Software and Other Performance Measures," *IEEE Trans. on Reliability*, vol. R-28, no. 3, pp.206-211, 1979.

[Grant 1992] A.S. Grant. "A context model needed for complex tasks," in *2nd Interdisciplinary workshop on Mental Models*, pp. 94-102, 1992.

[Henrion and Fischhoff 1986] M. Henrion and B. Fischhoff, "Assessing uncertainty in physical constants," *Americal Journal of Physics*, vol. 54, no. 9, pp.791-798, 1986.

[Hitch 1987] G.J. Hitch. "Working memory," in *Applying cognitive psychology to user interface design*, London, John Wiley, 1987.

[Jelinski and Moranda 1972] Z. Jelinski and P.B. Moranda. "Software Reliability Research," in *Statistical Computer Performance Evaluation*, pp. 465-484, New York, Academic Press, 1972.

[Knight and Leveson 1986a] J.C. Knight and N.G. Leveson. "An Empirical Study of Failure Probabilities in Multi-version Software," in *Proc. 16th IEEE Int. Symp. on Fault-Tolerant Computing (FTCS-16)*, pp. 165-170, Vienna, Austria, 1986a.

[Knight and Leveson 1986b] J.C. Knight and N.G. Leveson, "Experimental evaluation of the assumption of independence in multiversion software," *IEEE Trans Software Engineering*, vol. SE-12, no. 1, pp.96-109, 1986b.

[Lewis, Polson et al. 1990] C. Lewis, P. Polson, C. Wharton and R. J. "Testing a Walkthrough methodology for Theory-based design of Walk-up-and-use Interfaces," in *CHI-90*, pp. 235-241, ACM Press, 1990.

[Littlewood 1975] B. Littlewood. "A Reliability Model for Markov Structured Software," in *Proc. 1975 Int. Conf. on Reliable Software*, Los Angeles, IEEE . New York, 1975.

[Littlewood 1976] B. Littlewood. "A Semi-Markov Model for Software Reliability with Failure Costs," in *MRI Symp. Computer Software Engineering*, pp. 281-300, Polytechnic of New York, New York, Polytechnic Press, 1976.

[Littlewood 1979] B. Littlewood, "Software reliability model for modular program structure," *IEEE Trans Reliability*, vol. R-28, no. 3, pp.241-246, 1979.

[Littlewood 1981] B. Littlewood, "Stochastic Reliability Growth: A model for fault removal in computer programs and hardware designs," *IEEE Trans. on Reliability*, vol. R-30, pp.313-320, 1981.

[Littlewood 1988] B. Littlewood. "Forecasting software reliability," in *Software Reliability Modelling and Identification*, pp. 141-209, Heidelberg, Springer, 1988.

[Littlewood 1991] B. Littlewood. "Limits to evaluation of software dependability," in *Software Reliability and Metrics (Proceedings of 7th Annual CSR Conference, Garmisch-Partenkirchen)*, pp. 81-110, London, Elsevier, 1991.

[Littlewood and Miller 1989] B. Littlewood and D.R. Miller, "Conceptual modelling of coincident failures in multi-version software," *IEEE Trans on Software Engineering*, vol. SE-15, no. 12, pp.1596-1614, 1989.

[Littlewood and Strigini 1991] B. Littlewood and L. Strigini. *Validating ultra-high dependability for software-based systems*, Volume 3, Chapter 3, Part 1, PDCS, 1991.

[Littlewood and Verrall 1973] B. Littlewood and J.L. Verrall, "A Bayesian Reliability Growth Model for Computer Software," *J. Roy. Statist. Soc. C*, vol. 22, pp.332-346, 1973.

[Maiden and Sutcliffe 1992] N.A.M. Maiden and A.G. Sutcliffe, "Exploiting reusable specification through analogy," *Communications ACM*, vol. 35, no. 4, pp.55-64, 1992.

[Miller 1989] D. Miller. "The role of statistical modelling and inference in software quality assurance," in *Software Certification*, Barking, Elsevier Applied Science, 1989.

[MoD 1989] MoD. *Draft Interim Defence Standard 00-56, Requirements for the analysis of safety-critical software in defence equipment*, Ministry of Defence, London, 1989.

[Moser and Melliar-Smith 1990] L.E. Moser and P.M. Melliar-Smith, "Formal verification of safety-critical systems," *Software - Practice and Experience*, vol. 20, no. 8, pp.799-821, 1990.

[Musa 1975] J.D. Musa, "A Theory of Software Reliability and its Application," *IEEE Trans. on Software Engineering*, vol. SE-1, pp.312-327, 1975.

[Musa and Okumoto 1984] J.D. Musa and K. Okumoto. "A Logarithmic Poisson Execution Time Model for Software Reliability Measurement," in *Proc. Compsac 84*, pp. 230-238, Chicago, 1984.

[Norman 1988] D.A. Norman. *The Psychology of Everyday Things*, New York, Basic Books, 1988.

[Norman and Shallice 1986] D.A. Norman and T. Shallice. "Attention to action: willed and automatic control of behaviour," in *Consciousness and Self Regulation*, Plenum, 1986.

[Parnas, Schowan et al. 1990] D.L. Parnas, A.J.v. Schowan and S.P. Kwan, "Evaluation of safety-critical software," *Communications ACM*, vol. 33, no. 6, pp.636-648, 1990.

[Rassmussen 1986] J. Rassmussen. *Information Processing and Human Computer Interaction: An Approach to Cognitive Engineering*, North Holland, 1986.

[Ravden and Johnson 1989] S. Ravden and G. Johnson. *Evaluating Usability of*

Human Computer Interfaces, Ellis Harwood, 1989.

[Reason 1990] J.T. Reason. *Human Error,* Cambridge University Press, 1990.

[Reubenstein 1990] H.B. Reubenstein. *Automated Acquisition of Evolving Informal Descriptions.* 1990.

[Rouse 1981] W.B. Rouse, "Human-Computer Interaction in the Control of Dynamic systems," *ACM Computing Surveys,* vol. 13, no. 1, pp.71-99, 1981.

[Sanders and McCormick 1987] M.S. Sanders and E.J. McCormick. *Human Factors in Engineering and Design,* MacGraw-Hill, 1987.

[Shafer 1976] G. Shafer. *A Mathematical Theory of Evidence,* Princeton University Press, 1976.

[Siegrist 1988a] K. Siegrist, "Reliability of systems with Markov transfers of control," *IEEE Trans Software Engineering,* vol. SE-14, no. 7, pp.1049-1053, 1988a.

[Siegrist 1988b] K. Siegrist, "Reliability of systems with Markov transfers of control, II," *IEEE Trans Software Engineering,* vol. SE-14, no. 10, pp.1478-1480, 1988b.

[Sutcliffe and N.A.M. 1991] A.G. Sutcliffe and M. N.A.M., "Analogical software reuse: Empirical investigations of analogy based reuse and software engineering practices," *Acta Psychologica,* vol. 78, pp.173-197, 1991.

[Sutcliffe and Springett 1992] A.G. Sutcliffe and M.V. Springett. "From user's problems to design errors: Linking evaluation to improving design practice," in *HCI-92,* Cambridge Univ Press, 1992.

[Whiteside, Bennett et al. 1988] J. Whiteside, J. Bennett and K. Holzblatt. "Usability Engineering: Our experiences and evolution," in *Handbook of Human Computer Interaction,* North Holland, 1988.

[Wickens 1984] C. Wickens. *Engineering Psychology and Human Performance,* Columbus, Ohio, Merrill, 1984.

[Woods 1988] D.D. Woods. "Cognitive engineering in complex and dynamic worlds," in *Cognitive Engineering in Dynamic Worlds,* Academic Press, 1988.

[Wright and Monk 1989] P. Wright and A.F. Monk. "Evaluation for design," in *People and Computers,* Cambridge University Press, 1989.

[Xie 1991] M. Xie. *Software Reliability Modelling,* Singapore, World Scientific, 1991.

[Young and Barnard 1987] R.M. Young and P. Barnard. "The use of scenarios in human computer interaction research: Turbocharging the tortoise of cumulative science," in *Human Factors and the Graphics Interface,* pp. 291-296, New York, ACM Press, 1987.

[Young, Green et al. 1989] R.M. Young, T.R.G. Green and T. Simon. "Programmable user models for predictive evaluation of interface designs," in *CHI 89,* ACM, 1989.

A Modified Hazop Methodology For Safety Critical System Assessment

D J Burns, Dr R M Pitblado
DNV Technica, Lynton House, 7-12 Tavistock Square
London, WC1H9LT

Abstract

The concept of a Safety Lifecycle for Programmable Electronic System (PES) based systems has previously been suggested by the Working Group for an IEC draft standard dealing with safety-related software. The starting point in such a Lifecycle is Hazard Analysis which may be performed using a variety of accepted tools, including Fault Tree Analysis (FTA), Failure Modes, Effects and Criticality Analysis (FMECA) and Hazard and Operability studies (HAZOP). This paper describes work carried out to extend the standard FMECA and HAZOP approaches to suit the analysis of Programmable Electronic Systems, based on a number of case studies.

1   Introduction

The use of Programmable Electronic Systems (PES) in process control and emergency shutdown systems has grown considerably in the past 15 - 20 years. While generally superior in performance and reliability, compared to conventional controllers and hard-wired relay logic, PES can introduce novel hazards and several standards are currently being drafted at national and international level with the purpose of ensuring standardisation in all aspects of the specification, application and use of PES. Further information on current developments in this area is included in the proceedings of a recent symposium [SRD 92] and, for example, in [CE 92].

A PES is a computer-based system which controls, protects or monitors the operation of plant or machinery [HSE 87]. This HSE document also reiterates that when assessing the safety of PES, as in any system, the first stage is to identify the hazards and the chain of events that could lead to those hazards. This is called the Hazard Analysis. The estimation of the frequency with which the consequences of specified hazardous events occurs is called Risk Assessment, and is used to judge the acceptability of a system.

This paper is concerned chiefly with Hazard Analysis and Risk Assessment when PES are used to carry out critical safety functions and the use to these ends of a modified Hazard and Operability (HAZOP) methodology. In order to put the HAZOP in perspective, some other major factors, influences and definitions will be addressed in the first part of the paper, beginning with the concept of Safety Lifecycle, followed by a short discussion of software risks, some case histories and a description of the development of the HAZOP approach to Hazard Analysis, including the increasing awareness of the significance of human error.

2   The Safety Lifecycle

Several bodies have either published, or are developing, guidelines to enable the safe exploitation of PES technology. Problems arising from the specification, design and assessment of software for use in safety-related systems have been studied in [IEE 89]. In the UK, the Health and Safety Executive (HSE) published guidelines in 1987 under the general title "Programmable Electronic Systems in Safety-related Applications" [HSE 87]. These were developed taking into account work going on within other countries, after considerable research and discussion with industrial interests. At the international level, the International Electro-technical Commission (IEC) has issued two draft standards for systems [IEC 89a] and applications [IEC 89b] with the consultation period still under way.

The systems draft Standard is Part 1 of a planned series and a key feature is that each part will deal with specific aspects. It is proposed that future parts will cover:

- Hazard Analysis
- Risk Assessment
- System Safety Elements
- Safety Integrity Requirements
- System Integrity Levels
- System Validation
- Retro-fitting
- Documentation

Examples of the application sectors coming within the scope of the Standard include:

- process industries (emergency shut down systems, fire and gas detection systems, boiler controls)
- manufacturing industries (industrial robots, machine tools)
- transportation (railway signalling, braking systems, elevation/lifts)
- medical (electro-medical equipment)

During the development of the draft Standard, a Safety Lifecycle concept was developed reflecting some of the titles of future parts of the Standard. The envisaged lifecycle comprises the stages illustrated in Figure 1 [IEC 89a, b]. This paper will concentrate on the first step of the model : Hazard Analysis.

A hazard is a set of conditions that can lead to an accident under certain plant or environmental conditions, whereas the ensuing risk is a combination of the frequency, or probability, and the consequence of a specified hazardous event.

The hazard analysis is a structured study, the scope of which is determined by the potential consequence or risk in question and the system complexity. It may be carried out using, mainly, the following tools:

- fault tree analysis (FTA)
- failure modes, effects and criticality analysis (FMECA)
- hazard and operability study (HAZOP)

The objective of the hazard analysis is to identify all the hazards together with the events which could give rise to them, the safeguards in place and the consequences of their occurrence.

The application of the above tools to PES is addressed generally in [HSE 87]: the present paper describes a method of combining the features of FMECA and HAZOP.

3   Software Risks

A PES is the computerised part of a safety-related system, defined as a system that:

- implements, independently of any other system, the required safety functions necessary to achieve a safe state for the Equipment Under Control (EUC) or to maintain a safe state for the EUC.
- achieves, on its own or with other safety-related systems, the necessary level of safety integrity for the implementation of the required safety functions.

Safety integrity is defined as the likelihood of a safety-related system achieving its required functions under all the stated conditions within a stated period of time.

Safety-related systems can broadly be divided into two types: control systems and protection systems, although not all control systems will necessarily be designated as safety-related systems. It should be noted that a person could be part of a safety-related system (e.g. a manually deactivated ESD system) [Bell 91].

Software can increase risk in the following ways:

- causing the system to reach a hazardous state
- failing to detect and take corrective action to recover from a hazardous state
- failing to mitigate the damage after the occurrence of an accident or accident precursor.

The system can reach a hazardous state either because the software does not satisfy its specification or because the behaviour specified for it is not safe with regard to the overall operation of the system. As a guard against this type of behaviour, verification is essential to demonstrate that the execution or non-execution of the software should not result in a hazardous state [Leveson 91].

Failing to detect a hazardous system state and take corrective action against it may occur if the software has not been assigned to these tasks: the hazardous state may arise from a larger system independently of the behaviour of the software itself. Any such failure of the software is potentially safety-critical unless the larger system or the software or both have been designed to be safe, despite some types of software failure. For example, a plant shutdown is generally required within a specified time after detection of an unsafe state by the monitoring system: timing errors in the software could be safety-critical despite an overall correctly executed electro-mechanical function to trip the plant systems.

Failing to mitigate damage once an accident has occurred may be the result of plant areas being, incorrectly, untripped even though the monitoring systems indicate an unsafe condition there.

An important distinction is made in [Bell 91] between the terms Reliability and Safety Integrity within the context of the HSE guidelines [HSE 87]. Reliability addresses failures of a system due to hardware failures, that is, hardware failures resulting from various breakdown mechanisms that occur at random. Safety Integrity addresses both reliability and systematic failures that are due to errors that have been made at some stage in the specification, designs, construction, operation or maintenance of a system. Systematic failures are particularly important in the context of complex systems and are therefore particularly relevant to PES.

In the next section, some incidents previously analysed by DNV Technica [Pitblado 89] from computer controlled plants in the Netherlands are discussed.

4 Some Case Histories

At the time of the study addressed in the present paper, [Pitblado 88] a total of 17 incidents had been reported in the Netherlands over a 4 year period on computer controlled plants, which related to the computer system or to the human interaction with the computer. These failures led mainly to small and medium scale releases from the flare systems, but in one case led to plant damage and in another to a fireball. Table 1 shows a summary checklist indicating the number of failures in each category (hardware, software, human), from which it can be seen that human errors during operations were associated with 59% of the incidents. Errors were mainly due to inadequate, insufficient or incorrect information supplied to the operators (59%) and a failure to correctly follow procedures (47%). Human errors in design were involved in 29% of incidents. Hardware and software failures were less prominent.

From a model of human behaviour, a more detailed breakdown of the causes of the 17 incidents is given in Table 2. It can be seen that poor information provision, whether incorrect, hidden or not available, derives from a number of sources. For example, the quality of procedures, supervision and checking may be insufficient to enable errors to be identified and recovered in installation or maintenance. Over reliance on the computer when carrying out procedures could, however, reflect inadequate understanding by operators of the functions performed by the computer.

Some recent information from vendors and users is included in [SRD 92] which relates further case histories.

With the guidelines of the ongoing standards and the experience to date in mind, a methodology for hazard identification is described in the next section, combining the approaches of FMECA and HAZOP.

5 HAZOP Study for PES Hazard Identification

Conventional HAZOP procedures [CIA 90] are regarded as insufficient to address many potential problems of PES. They can, however, be made effective with a number of extensions, the total comprising three stages:

1. Initial "conventional" HAZOP
2. PES HAZOP
3. Human Factors HAZOP

The timing of such a sequence of HAZOPs is important in making the project both worthwhile and cost effective [Ford 90]. Successful studies have been performed on systems in both the design and operating stages, however, modifications are much easier and less expensive to implement in the early stages of the project. On the other hand, a sufficient level of detail must be defined for both the control system and process prior to the study:

5.1 Initial HAZOP

The initial HAZOP would be conducted using normal procedures for the whole system including the control system, but not at the level of detail to be addressed in the subsequent PES HAZOP.

5.1.1 Documentation

Normal HAZOP documentation requirements consist of Process Flow Diagrams, P&IDs or EFDs, site layout drawings, equipment detail drawings, supplemented as necessary with electrical one-line diagrams, operating instructions and trip system definitions.

5.1.2 Study Team

The composition of the team depends on the type of application, but should always use a member who has operating experience and a member who is closely familiar with the design. Other members are made up of other speciality disciplines and usually number between two or four. The study is chaired by a person who is experienced in the HAZOP technique, but not necessarily an expert on the process being reviewed.

5.1.3 Methodology

A list of variables and guidewords is used to stimulate the creative thinking of the study team, and which define the operation of the system. The list of basic variables includes:

> Pressure
> Temperature
> Flow
> Level
> Composition etc.

While the list of guidewords includes:

NO OR NONE	COMPLETE NEGATION OF DESIGN INTENT (e.g. No Flow)
MORE	QUANTITATIVE INCREASE FROM DESIGN INTENT (e.g. More Pressure)
LESS	QUANTITATIVE DECREASE FROM DESIGN INTENT (e.g. Less Pressure)
REVERSE	OPPOSITE OF DESIGN INTENT (e.g. Reverse Flow)
PART OF	QUALITATIVE DECREASE FROM DESIGN INTENT (e.g. Composition Change)
AS WELL AS	QUALITATIVE INCREASE FROM DESIGN INTENT (e.g. Composition)
OTHER THAN	COMPLETE SUBSTITUTION (e.g. Relief)

The meeting is minuted by a HAZOP Recorder who notes the causes and effects of identified deviations from the design intent, the safeguards in place, the need for design or operational checks or enhancement actions, who should address them, and by which date. DNV Technica normally uses purpose-designed HAZOP software to generate standard forms and data base information for subsequent tracking and follow-up. We would recommend that the conventional HAZOP be supplemented by a PES HAZOP and a Human Factors HAZOP.

5.2   PES HAZOP

5.2.1   Documentation

The conventional HAZOP documentation is supplemented to enable a review of computer control/PLC systems. The logic systems need to be as completely described as the PFD and P&ID do for process systems and hardware. Several types of documentation are available:

- cause and effect charts
- ladder diagrams
- logic diagrams
- vendor diagrams

Of the above types, the logic diagram has been found most easily understandable by the wide range of personnel involved in the study. In this diagram, input signals are displayed on the left hand side of the page and output signals on the right. The space in between is occupied by a representation of the software using the symbols for OR and AND gates, signal invertors, timers etc. This is not a standard design document and will be produced only if included in the project specification.

5.2.2   Study Team

The composition of the HAZOP team for a PES study is different from that for a traditional chemical process HAZOP. Besides the attendance of an experienced HAZOP leader, preferably involved in the initial HAZOP, the following should be present:

- control design engineer (possibly from vendor)
- instrumentation technician
- process engineer
- operating staff member
- corporate control engineer (optional)

For a period in any stage of development, the control engineers who designed the control strategies for the system should be included. At the beginning of each study they should explain the design intention of the logic contained in the section.

Instrumentation technicians should be included to provide historical accounts of maintenance problems and typical failure modes. They are also valuable in the development of practical recommendations. A representative from process engineering will be able to comment on the current operating and safety issues associated with the control of the process and approve recommendations that will effect the operations of the unit. For projects that involve the corporate control group, or require approval of the group, a representative should be involved in the HAZOP. Their knowledge of corporate standards such as the use of by-passes on Emergency Shut Down devices or control valves as ESD valves, will be important in both the identification and recommendation phases.

5.2.3 Methodology

A modification of the Failure Modes Effects and Criticality Analysis (FMECA) [Mil Std 1629A] is used, using HAZOP-type guidewords. This approach has met with good results when applied by DNV Technica in the past. Its operation is similar to a conventional HAZOP, but it uses the following guidewords:

VARIABLE	GUIDEWORD
SIGNAL	NO
ACTION	MORE
	LESS
	WRONG

A sample log from a study session is shown in Figure 2.

5.3 Human Factors HAZOP

5.3.1 Documentation

As with the review of the PES, additional documentation should be assembled, including:

- definition of all important VDU displays
- the number of VDU screens and their function
- hardcopy devices
- control room layout and ergonomics (lighting, noise level, distractions, etc)
- relationship of computer to hardwired systems and alarms
- access times to data
- communications
- the system interface and potential for manual override, disabling of automatic functions, set-point variations, etc
- job role and task definitions
- procedures

5.3.2 Methodology

The guidewords used to explore human factors deviations are similar to those used in the PES HAZOP, but now apply to human actions rather than control systems.

VARIABLE	GUIDEWORD
INFORMATION	NO
	MORE
	LESS
ACTION	NO
	WRONG

A full explanation as to the meaning of these combinations is given in [Bellamy 88]. A sample log from Human Factors HAZOP is given in Figure 3.

The variable INFORMATION applies to information available from displays, procedures, previous training, experience, communications or any other source the operator may use. The variable ACTION refers to the operator response. Errors in ACTION may be in terms of incorrect selection or incorrect execution of a response.

5.4 Main Findings [Pitblado 89]

Applying the above techniques to distributed control systems has resulted in the identification of numerous safety and operability problems, and provided possible solutions for most of them. The technique covers issues such as signal failure, alarm philosophy and its application, fail positions, equipment failures, maintenance, and the design of the operator interface. Major findings occurred in the following areas:

- Failure to meet the design intent
- Dependency between redundant instruments
- Human Factors
- System maintainability
- Compromised emergency shutdown system

A study of a control system just before start-up identified numerous sections of the code that failed to meet the design intent. Without the HAZOP review, these problems would have gone undetected until the system was on-line where troubleshooting would have been difficult and expensive. The team also identified areas where a single failure could defeat attempts to provide redundant protection of the system. This occurred when both instruments were wired to the same I/O card. The teams also addressed issues concerning the amount and type of information provided to the operator, the ease of interpretation and the ability to easily perform the correct responsive action. Instruments that could not be maintained without initiating a process shutdown were addressed through recommendations to provide a bypass, which would allow routine maintenance and calibration to be easily accomplished. The topic of shutdown philosophy and implementation is a critical area covered by the HAZOP. Findings in this area included errors that, under certain conditions, prevented the "STOP" push button from actually shutting the system down and conditions that could lead to the system starting without all of the required permissions. These errors were identified following numerous other logic and system reviews.

6. Conclusions

The requirement for a Hazard Analysis review of Programmable Electronic Systems has been identified as the first step in a Safety Lifecycle for PES systems. The Safety

Lifecycle concept has grown from current international groups working on the development of guidelines for PES.

An approach which combines the strengths of FMECA and HAZOP has been found to be very successful in meeting this requirement, as applied to a number of process plants. The methodology comprises a conventional HAZOP, a PES HAZOP and a Human Factors HAZOP.

7. References

[HSE 87]   Health and Safety Executive: Programmable Electronic Systems in Safety Related Applications 1. An Introductory Guide 2. General Technical Guidelines, HMSO, 1987

[Pitblado 88]   Pitblado R M, Bellamy L, Geyer T: Safety Assessment of Computer Controlled Plants, Technica Limited, London, 1988

[Bellamy 88]   Bellamy L J, Geyer T A W: in "Human Factors and Decision Making - Their Influence on Safety and Reliability" Safety and Reliability Symposium, Manchester 19-20 October 1988

[IEE 89]   Software in Safety-related Systems: The Institute of Electrical Engineers and the British Computer Society 1989

[IEC 89a]   IEC draft standard: 65A (Secretariat) 123 "Functional Safety of Programmable Electronic Systems. Generic Aspects Part 1

[IEC 89b]   IEC draft standard.   65A (Secretariat) 122:   "Software for Computers in the Application of Industrial Safety-related Systems"

[Pitblado 89]   Pitblado R M, Bellamy L, Geyer T: Safety Assessment of Computer Controlled Process Plants 6th International Symposium "Loss Prevention and Safety Promotion in the Process Industries", Oslo, Norway June 19-22, 1989

[Ford 90]   Ford K A, Brown W H: Innovative Applications of the HAZOP Technique, AI Ch E Spring National Meeting, Orlando, Florida, March 20, 1990

[CIA 90]   Chemical Industries Association: A Guide to Hazard and Operability Studies CIA 1990

[Bell 91]   .   Bell R, Smith: Functional Safety of Programmable Electronic Systems Management and Engineering of Fire Safety and Loss Prevention Onshore and Offshore bHr Group, Aberdeen, February 1991

[Leveson 91]   Leveson N G, Cha S G, Shimeall T J: Safety Verification of ADA Programs Using Software Fault Trees. IEEE Software July 1991, 48 - 59

[SRD 92]   Reliability of Programmable Electronic Systems. The SRD Association. Proceedings of Symposium, Risley, 18 March 1992 - SRDA - R6.

[CE 92]   Sawyer P: Software for Safety? The Chemical Engineer 10
September 1992 pp 32-34.

[MIL-STD-1629A]   Procedures for Performing a Failure Modes, Effects and Criticality
Analysis

FIGURE 1: SAFETY LIFECYCLE MODEL

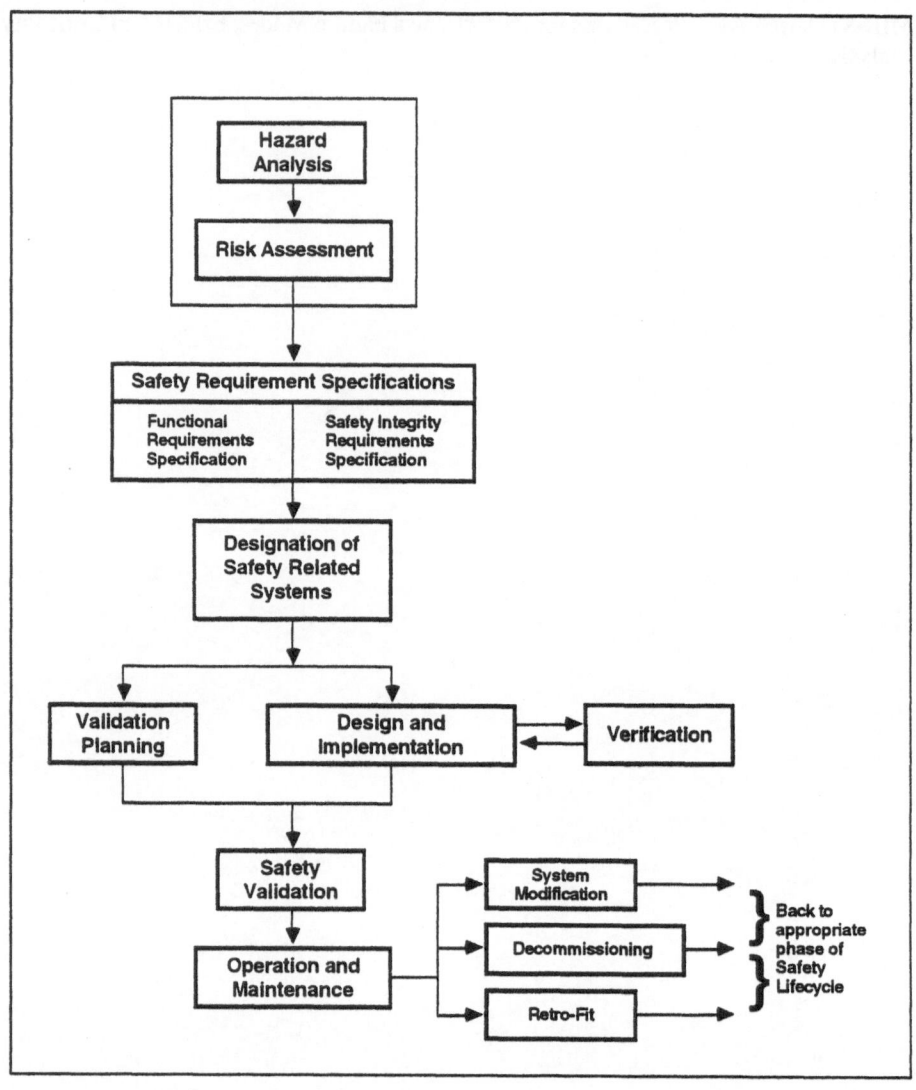

FIGURE 2: SAMPLE LOGSHEET FROM A PES HAZOP

SYSTEM:PRC 570		SUB-SYSTEM FIC 570	
DEVIATION	POSSIBLE CAUSES	CONSEQUENCES	RECOMMENDATIONS OR QUESTIONS
NO SIGNAL LESS SIGNAL	Transmitter failure Sensor reads high	FV 570 opens wide; reactor pressure falls	R203: Add independent low pressure alarm
MORE SIGNAL	Sensor reads low	High pressure - trip fires	
NO ACTION	Valve jammed	Loss of control	R204: Consider providing block and bypass
MORE ACTION	Air failure	Reaction pressure falls	R205: Perform systematic review of air failure problems
High and low alarms are provided from flow sensor; high alarm only from PRT 570			
FMA LOGSHEET NO. 057	PROJECT C970	LOGIC DIA. NO. R 23 SPC 570 2	SHEET 1 TO 7

FIGURE 3: SAMPLE LOGSHEET FOR HUMAN FACTORS HAZOP OF FUNCTIONS: NO INFORMATION

FUNCTION: MONITOR FLARE LINE PRESSURE			
DEVIATION	POSSIBLE CAUSES	CONSEQUENCES	FURTHER ANALYSIS
No information	Information not perceived or attentional problem or lost from memory	Fail to identify high pressure in flare line - no recovery response	Alarm available in displays?
	- No alarm provided		Installation/mainten- ance procedures for this system
	- Operator not monitoring display		
	- Operator forgets		Related failures/work demands/priorities training/manning

TABLE 1:   FAILURE CONTRIBUTORS TO 17 INCIDENTS

FAILURE CATEGORY	DETAILED FAILURE	FACTOR INVOLVED IN	
		NUMBER OF INCIDENTS	% OF INCIDENTS
HARDWARE	Computer Hardware	3	18
	Connection Hardware		
	- Electronic	0	0
	- Pneumatic	0	0
	- Electrical	1	6
	Protective System Hardware	0	0
	Equipment Hardware	5	29
SOFTWARE	System Software (Manufacturer's Shell)	1	6
	Site software implementation (i.e. Software written for the process plant and installed during and after implementation)	2	12
HUMAN	Error Context		
	- Design	5	29
	- Installation	2	12
	- Commissioning/ Testing	1	6
	- Operating	10	59
	- Maintenance	2	12
	Error Type		
	- Failure to follow procedure (correctly)	8	47
	- Recognition failure, given adequate supply of information	2	12
	- Error due to inadequate/insufficient/incorrect information supplied to person(s) involved	10	59

TABLE 2:   BREAKDOWN OF CAUSES OF THE 17 INCIDENTS

HUMAN SOFTWARE ERRORS	INCIDENT NUMBER CODE
Interface does not display actual plant status	1, 6, 8, 16
Installation error leads to incorrect information	3, 8
Alarm set incorrectly	4
No alarm (maintenance error)	4, 5
No alarm (design)	4, 5
Operator misses information due to overload	2, 13, 15, 17
No independent means of cross checking provided	1, 3, 6, 16
Operator fails to cross check	8
Trip disabled/manual override	1, 8, 11
Over-reliance on computer	9, 11, 14, (15)
Inadequate knowledge	(3), 11
Failure to update operator's information	12, 17
Incorrect control signal (maintenance error)	10
Design error: Plant	4, 5, 8, 17
Design error: Computer control system	7
HARDWARE FAILURES	
Equipment hardware	2, 5, 13, 14, 15
Computer hardware	7, (11), 16
Connection hardware (electrical)	6

(Note: Incident numbers in parentheses indicate that there was not enough information to allocate to that failure category with certainty).

Applying Security Techniques to Achieving Safety

Dr. David F.C. Brewer

Gamma Secure Systems Limited, Diamond House

149 Frimley Road, Camberley, Surrey GU15 2PS

1   Introduction

In October 1986, the Centre for Software Reliability sponsored a symposium on Safety and Security. Held in Glasgow, the aim of this symposium was to explore the proposition that safety and security were duals and that there would be much to be gained by the one adopting the knowledge, understanding, tools and techniques of the other, and vice versa. The symposium [Anderson 89] concluded that "the differences between the topics of Reliability, Safety and Security are more than sheer semantics" but that "there are many similarities which we would be wise to explore".

It was stated at the symposium that "a computing system may be termed *safe* if it will not cause death or injury" and that "it may be termed *secure* if it will not betray any secrets". Whilst some people prefer definitions such as "security is the prevention of fraud or error", one thing that can be agreed is that security concerns the protection of information against some threat, whereas safety concerns the protection of people against some hazard.

This is a useful distinction: reliability of a computer system will be a security problem if its failure results in the unauthorised disclosure, modification or withholding of information, or a safety problem if its failure results in death or injury.

Now, one would imagine that the techniques necessary to ensure the reliability of a computer system would be the same, regardless of the reasons for why one would want to make a computer system highly reliable in the first place. Yet, as evidenced at the symposium, the views of the safety engineer and the security engineer were quite different, with the security engineer showing a preference for fault prevention techniques (i.e. techniques designed to prevent something from happening) and the safety engineer a preference for fault tolerant techniques (i.e. techniques for countering something that has happened). It is important to question this, since if the difference is more fundamental than mere predilection the relevance of security techniques to achieving safety might be rather tenuous.

In this paper we will describe a number of security techniques. These techniques were not developed at the same time, nor were they developed in a systematic manner, although we will present them as if they were. However, it is useful to know something of the chronological order of development, at least to gain some sense of where the world of security engineering is going. To this end we will first give a brief account of the history of computer security evolution.

In conclusion we will notice that security engineers have come to appreciate that security must be achieved through a combination of both fault preventative and fault tolerant techniques. This should be a result of utmost importance to the safety engineer. Safety engineers are now encouraged to use fault prevention techniques (e.g. through the publication of [MOD 91]). To maximise the benefit of that approach the safety engineer must adopt the security principles of the "reference monitor concept" and security policy modelling; but in doing so, he (she) must not forget the more traditional safety engineering approach of fault tolerance.

2 History of Computer Security Evolution

2.1 The US Initiative

In October 1967, the US Department of Defense (DoD) commissioned a task force to address computer security safeguards for the protection of classified information in remote-access, resource-sharing computer systems. This led to the publication of a set of security evaluation criteria (the "Orange Book", [DOD 85a]) and the establishment of the National Computer Security Center (NCSC) to carry out product evaluations in accordance with the criteria. The purpose of the "Orange Book" is to prescribe given functionality at given assurance levels to meet given levels of risk. Operating systems could then be built against the "Orange Book" knowing that they would then meet generic US government requirements.

The scientific basis of the "Orange Book" rests upon the *reference monitor concept* [Anderson 72] and the Bell-LaPadula security policy model [Bell 76], and indeed this basis has led to the development of commercially available (off-the-shelf) operating systems which can *enforce* the US DoD rules for processing classified information.

Naturally, within the two decades that separated the beginning of the US computer security initiative and the emergence of secure operating systems in the market place, computer technology evolved significantly. As a consequence the NCSC was forced to develop interpretations of the "Orange Book" to deal with computer networks and databases, and more recently "window" applications. In these cases, the scientific basis of the "Orange Book" remained unchanged; it is only the way in which it is applied that is different.

However, in 1987, two Americans, David Clark and David Wilson, published a comparison of military and commercial security policies [Clark 87]. They showed that the security of commercial computer systems (e.g. banking systems) could be described by a set of "security objectives" analogous to Bell and LaPadula's security axioms (see Annex A), yet not all of them could be enforced through utilisation of the reference monitor concept (indeed some security mechanisms cannot be implemented in either software or hardware and must rely on human intervention). This provided a direct challenge to the work of Bell and LaPadula. Radically different security policies were then identified, modelled and published: for example the Chinese Wall Security Policy [Brewer 89] which was based on the UK Financial Services Act. In 1991, the US DoD undertook a study of mechanisms used to provide information integrity which identified many more, even more radially different security policies, drawn from common US military practice and commercial practice worldwide. Nevertheless, despite recognition that there is no longer any general purpose security policy model (as was once thought with the

247

Bell-LaPadula model, certainly in 1986), it is possible that most policies, if not all, can be described in terms of the lattice-framework of the Bell-LaPadula model [Bell 91].

2.2 The European Initiative

UK interest in computer security started with the appointment in 1982 of the Communications-Electronics Security Group (CESG) as the National Authority for all UK government systems processing classified information. CESG established a set of criteria, appropriate to the evaluation of complete systems (including physical, personnel and administrative aspects), and an evaluation scheme which became operative in 1984. In 1987, the Department of Trade and Industry (DTI) established its Commercial Computer Security Centre (CCSC). The DTI's objective was to promote the interests of UK industry and saw a way of doing this through the development of internationally recognised security evaluation criteria, appropriate to products and all market sectors. The CCSC recognised that the way to do this was to generalise the "Orange Book" by unbundling its security functionality requirements from its assurance requirements; a view shared by CESG and indeed by many other countries, notably France, Germany, the Netherlands and Canada. In particular it is necessary to be able to build functionally simple products (such as cryptographic devices) with very high assurance and more complex products (such as databases) with correspondingly lower assurance.

Another DTI objective was to establish an open "commercial" evaluation scheme. The NCSC scheme was only open to US vendors, the CESG scheme was open only to UK government. Thus a UK vendor was unable to commission an NCSC evaluation and could only commission a CESG evaluation with the assistance of a UK government sponsor. This situation changed in 1991 with the establishment of a single UK IT Security Evaluation and Certification Scheme jointly managed by CESG and the DTI. (We should note in passing that the idea of third party independent evaluation is important to security evaluation as it provides independent "proof" that the manufacturer has not, either deliberately or accidentally, introduced a security weakness into the product or system under evaluation.)

Also in 1991, the Commission of European Communities published the Information Technology Security Evaluation Criteria (ITSEC, [CEC91]) - harmonised security evaluation criteria of France, Germany, the Netherlands, CESG and the CCSC. The four nations are now engaged in developing the Information Technology Security Evaluation Manual (ITSEM), which will provide the technical basis for mutual recognition of evaluation certificates; the idea here being that the result of an evaluation in France, say, is recognised in Germany just as if the Germans had conducted the evaluation. The ITSEM is based on the principles set out in the European Standard EN45001 (ISO Guide 25) for achieving mutual recognition of test results. Indeed a worldwide framework has already been established to do this for other forms of testing (e.g. chemical analyses), and the four nations simply aim to take advantage of this for security evaluation. The UK

Scheme is founded on these principles and serves as a direct model for future schemes.

2.3 The International Initiative

In 1990 the US Computer Security Act devolved powers to the US National Institute for Standards and Technology (NIST) similar to those enjoyed by the DTI in the UK. Quite understandably NIST has adopted a similar relationship with the NCSC as the DTI has with CESG. Spurred on by the development of the ITSEC and the ITSEM, NIST started developing a new set of US criteria - the Federal Criteria.

The Canadians meanwhile published their own criteria. These criteria also separate functionality from assurance, but identify a number of useful security functional building blocks, with advice on how to put them together.

The US regard the ITSEC as the first significant step towards an international standard and the Canadian criteria as the second; they hope that their own Federal Criteria will become, if not that international standard, the third and final step towards that standard.

3   The Reference Monitor Concept

The *reference monitor concept*, first identified in the Anderson Report [Anderson 72], is an essential ingredient of any secure computer system. Its implementation, the *reference validation mechanism*, permits the *enforcement* of particular types of security property by some small component of the system. In other words, one good (correct) program, written in a certain way, can force all other programs to obey the rules of security, regardless of whether they themselves are correct programs or not. This concept is of fundamental importance to the security engineers' fault preventative approach. In particular it characterises his (her) approach to "architectural design".

The idea is simply that enforceable security rules can be enforced system-wide by a single component which is:

- tamperproof
- always invoked
- small enough to be subject to analysis and tests, the completeness of which can be assured.

We will illustrate this concept by way of an example.

Figure 1 shows a schematic of a modern-day purpose built computer in which the processor and memory chips responsible for executing application software and security software are separated by an Ethernet LAN. All of this logic would be housed in a single box which could be made as tamper resistant as desired. Its architecture, simplified so as not to overburden this discussion, nevertheless illustrates a number of important principles:

- The fixed disc is only directly accessible by the security processor. This means that neither the user nor the application software can access the disc without the knowledge of the security processor. In turn this means that the security processor can grant or deny access to the disc in accordance with some security rules, for example axioms (2a)

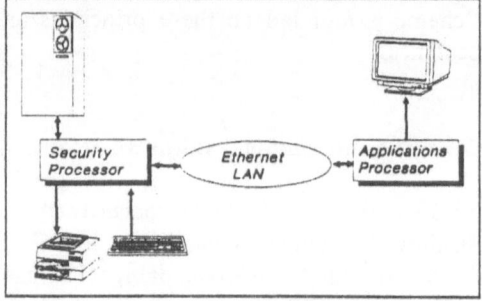

Figure 1: Application of the Reference Monitor concept to hardware architecture design

and (2b) described in Annex A. Moreover, the security processor can be programmed to maintain a log of all transactions, including attempted security violations on the fixed disc.
- It is not possible for the applications software to modify the security software without its cooperation since its memory is not directly accessible from the applications processor.
- In order to uphold the security axioms, we only need to verify that the software in the security processor only does what it is supposed to do and does not do anything that it is not supposed to do. Given this, the system will uphold the security rules no matter what the application software does.

4   Formal Models

The work of Bell and LaPadula is of fundamental importance to the development of computer security, not only because it precipitated the events that were then to follow, but because it demonstrates how some critical property of a computer system (in this case a security property) can be cast as an axiomatic model. If a computer system can be built which can be formally demonstrated to satisfy the axioms of such a model, referred to as a security policy model, then we have (probably) the greatest level of confidence ever achievable in that our security needs are met.

The essence of the Bell-LaPadula model is given in Annex A. It models the interaction between a user and the computer resources, specifically information, that he (she) wishes to access and manipulate, with access governed by the rules of some national security policy. The formulation of the Chinese Wall Security Policy [Brewer 89] is similar to that of Bell and LaPadula; the security rules are, however, very different, being formulated in accordance with the UK Financial Services Act.

In the Bell-LaPadula case, each user of the computer system is given a "clearance" which entitles that user to access information up to a certain "classification". This clearance does not normally change with time, the general exception being the revocation of all rights of access when a user leaves the organisation. In the Chinese Wall case, the user's clearance changes dynamically as a result of accessing the computer: once a user has chosen to access information on a particular company, he (she) is prevented from accessing any information belonging to a competitor (so as to inhibit the opportunity for "insider dealing").

The important difference between these two models is that they are models of different things; it is the objects of the model, an operating system processing classified information on the one hand and a financial services department on the other, that are different - not the models per se. Indeed, Bell and LaPadula's only objective was to develop a secure operating system; they were not concerned with what happened outside of the computer system (e.g. preventing a user from posting a copy of a top secret document to a spy). Moreover, the Bell-LaPadula model does not faithfully reflect all activities traditionally practised in a secure non-automated office environment.

Thus, notwithstanding the power of the policy modelling approach, care must be taken to model all the relevant aspects of the real world. However, if something important is missed, then it might be possible to counter it when the problem is viewed in the wider system context, described next.

5   Systems

We will regard a "system" as a specific IT installation, with a particular purpose and operational environment, for example a network of automated teller machines (ATMs). In this case, the IT installation comprises the ATMs that we see in our streets, the mainframe computers that control them and the networks that connect them. The environment consists of physical things (such as the buildings in which the ATMs and other components are situated) and people (the customers, the bankers and, of course, the robbers).

CESG recognised that responsibility for ensuring security of a complete system may fall to a variety of organisations. This is the case in our example. A given branch would only be responsible for the security of its own ATM. All other components would be the responsibility of the branches in which those other components were situated. Indeed for a large branch, responsibility for the security of the ATM might fall to one department, security of a mainframe controlling computer (if it had one) to another department and responsibility for overall site security to a third.

The existence of these different areas of responsibility partitions the overall system into a number of separate domains of responsibility. For example, CESG refers to the domain, which is the responsibility of a given security officer, as the local security environment (LSE) and all other domains (which are not that person's responsibility) as the global security environment (GSE). CESG further recognised the distinction between the non-technical countermeasures (i.e. physical, personnel and administrative procedures) and the technical countermeasures (e.g. cryptographic devices, computers, printers, networks and so on). CESG chose to regard only the non-technical countermeasures to be part of the LSE and allocated the technical countermeasures to a third domain, the electronic security environment (ESE).

In developing a system security policy, the security engineer must first identify these three domains. He (she) then identifies the most appropriate

251

countermeasures to protect the system's information assets from the perceived threats and allocates them to one or other of the three domains. Generally, the security engineer makes assertions about the characteristics of countermeasures which should already exist (e.g. in the GSE) and identifies the requirements of those which need to be developed and installed. An obvious point is that the security engineer will make use of the "principle of defence in depth". Staff whose job it is to program and maintain the ATM controlling mainframe computers will have their identity checked on entry to the bank, on entry to the computer room as well as by the computer itself when they log on at a terminal to start work. Moreover, it is not always possible to counter a given threat with a single countermeasure. In these cases a number of mutually supportive countermeasures need to be identified. Quite often these supportive countermeasures are placed in different domains with a view to minimise the opportunity for introducing a point of "common mode failure".

Viewed in this context, it is unnecessary to place infinite reliance in any particular component, each countermeasure being designed to uphold security even in the event of the failure of another countermeasure. Clearly, this is a fault tolerant approach.

6 Assurance

The objective of the security engineer is to gain assurance, i.e. confidence that the security objectives of a system are met. The ITSEC tells us that to do this we must:

- select the right countermeasures (the ITSEC functionality criteria)
- position them so that the security objectives will be met, regardless of how the system might be attacked (the ITSEC effectiveness criteria)
- ensure that each countermeasure does what it is supposed to do and does not do anything that it is not supposed to do (the ITSEC correctness criteria).

Of importance is that one cannot be 100% confident in either "effectiveness" or "correctness". However, we may gain confidence in effectiveness by the introduction of additional countermeasures, arranged to counter the defects of the existing countermeasures. Thus "effectiveness" represents the security engineers' fault tolerant approach to security. Confidence in correctness is gained by using evermore exacting and rigorous approaches to development, not just in the development process but in the development environment as well (for example by using formal specification methods like Z, supported by formal code analysis tools like SPADE, and tool based configuration management methods); it represents the fault preventative component of the security engineers' approach.

Of equal importance is how much confidence do we need anyway. The answer will depend on severity of the threat (perhaps measured in terms of the opportunity, expertise and resource that an attacker may bring to bear (see ITSEC)), the attractiveness of the target (perhaps measured in terms of financial gain to the attacker) and the consequences of a successful attack (perhaps measured in terms of financial loss to the owner of the system).

The marriage of how much confidence do we want (risk assessment) to the determination of how much confidence do we have (security evaluation) is the final part of our story.

7   Criteria and Risk Assessment

The "Orange Book" defines six classes of operating system (classes C1, C2, B1, B2, B3 and A1 respectively), each possessing a greater level of trust than its predecessor. It requires developers:

- to build their products using particular engineering techniques (both in terms of the development process, for example when formal methods are to be used, and in terms of hardware and software architecture, for example as illustrated in Figure 1) and
- to ensure that their products meet certain (security) functionality requirements.

The level of trust afforded by each class of operating system is prescribed in a separate NCSC document, the "Yellow Book" [DOD 85b], which is an example of a basic risk assessment method (the security engineer's equivalent of the safety engineer's hazard analysis). Just to take two extreme examples we would find:

- For class C1 operating systems, all users must be cleared to at least the highest classification of any information held on the system. For example, if UNCLASSIFIED and SECRET information is held, then all users must be cleared to access SECRET information. The users may, however, be formed into groups (e.g. different user departments) and the system will prohibit one group from accessing another group's data. Application software of unknown pedigree can be used.
- For class A1 operating systems, information ranging from UNCLASSIFIED to TOP SECRET may be held on the system, with access permitted by uncleared personnel, provided that the pedigree of all application software is known.

However, for certain situations involving most highly classified information the "Yellow Book" reveals that even a class A1 operating system is not good enough, requiring some level of assurance "beyond" A1. The way around this, clearly is to use a combination of fault prevention and fault tolerant approaches.

Precisely how this is done is the subject of current research. Of particular interest, however, is that the "Yellow Book" concept has been extended by the introduction of other factors, such as the number of users, the types of terminal used and the quantity of data processed. Given these extensions it is possible to show that a stand-alone computer system which requires "beyond A1" assurance by the "Yellow Book" can be implemented by a distributed "fault-tolerant" computer system using available off-the-shelf technology. This is a promising avenue of research, and one which might benefit from a greater understanding of the safety engineers hazard analysis techniques.

9   Conclusion

In conclusion, we note that the fault preventative approach provides high assurance in a single component. It is most effective when based on a formal security policy model and implemented using the reference monitor concept. Weaknesses in the security policy model can, however, be overcome by combining the fault preventative approach with a fault tolerant approach, indicating that the path to greater assurance requires a combination of the two methods.

Safety engineers are now encouraged to use fault prevention techniques (e.g. through the publication of [MOD 91]), in particular the use of formal methods to the development of safety-critical software.   To maximise the benefit of this approach the safety engineer would be wise to adopt the security principles of the "reference monitor concept" and security policy modelling. However, as in the case of security, the safety engineer would also be wise to utilise the more traditional safety engineering approach of fault tolerance in harmony with the new fault preventative approach.

References

[Anderson 89]   Anderson T: "Safe and Secure Computing Systems." Blackwell Scientific Publications, London, 1989.

[Anderson 72]   Anderson J P: "Computer Security Technology Planning Study." ESD-TR-73-51, vol I, AD-758 206, ESD/AFSC, Hanscom AFB, Bedford, Mass., October 1972.

[Bell 76]   Bell D E and LaPadula L J: "Secure Computer Systems: Unified Exposition and Multics Interpretation." ESD-TR-75-306 MTR 2997 Rev 1, The MITRE Corporation, USA, March 1976.

[Bell 91]   Bell D E: "Putting Policy Commonalities to Work." Proceedings of the 14th National Computer Security Conference, NIST/NCSC, USA, 1991 pp 456-471.

[Brewer 89]   Brewer D F C and Nash M J: "The Chinese Wall Security Policy." Proceeding of the IEEE Symposium on Security and Privacy, 1989 pp 206-214.

[CEC 91]   Commission of European Communities: "Information Technology Security Evaluation Criteria - Provisional Harmonised Criteria." Commission of European Communities, Luxembourg, June 1991.

[CEC 92]   Commission of European Communities: "Information Technology Security Evaluation Manual." V 0.2, Commission of European Communities, Luxembourg, March 1992.

[Clark 87]   Clark D and Wilson D R: "A Comparison of Commercial and Military Computer Security Policies." Proceedings of the 1987 IEEE Symposium on Research in Security and Privacy, 1987 pp 184-194.

[DOD 85a]   US Department of Defense: "Trusted Computer System Evaluation Criteria." DOD 5200.28-STD, Department of Defense, USA, December 1985.

[DOD 85b]   US Department of Defense: "Guidance for Applying the Department of Defense Trusted Computer System Evaluation Criteria in Specific Environments." CSC-STD-003-85, Department of Defense, USA, June 1985.

[MOD 91]   Ministry of Defence: "Interim Defence Standard 00-55/Issue 1 - The Procurement of Safety Critical Software in Defence Equipment." MOD, April 1991.

Annex A - The Essence of the Bell-LaPadula Model

Prior to 1987, the subject of military confidentiality dominated computer security literature. The basic problem can be stated as follows. Information is assigned a level of importance, called a "classification" which reflects the damage to the nation that might result if that information was revealed to an enemy of the state. In the UK there are five classifications: UNCLASSIFIED, RESTRICTED, CONFIDENTIAL, SECRET and TOP SECRET, representing an increasing sequence of damage from "no damage" to "grave damage". People permitted to access this information are granted a "clearance" which may be mapped onto the set of classifications to give the highest classification to which that person may be granted access. Security is then (partially) defined by the (semi-formal) axiomatic statement:

$$S \text{ may access } O \text{ if and only if } CLR_S \text{ dom } CLASS_O \tag{1}$$

where CLR_S represents the clearance of some "subject", S, (i.e. a person or software program acting on his (her) behalf) expressed in terms of a classification, $CLASS_O$ represents the classification of some "object" (e.g. a disc file), O, which S wishes to read, and **dom** is the dominance operator used to compare elements of a partially ordered set. The partially order set, in this case, is the set of classifications {UNCLASSIFIED, ..., TOP SECRET}. CLR_S and $CLASS_O$ are short hand notations for two partial functions which map the sets of all possible users of a computer system (which strictly speaking means everyone in the world) and all computer files (which again streaking includes all magnetic media which could be accessed by the system) onto the set of classifications.

Axiom (1) is only a partial statement of security since a subject S_1 with a SECRET clearance could open two files, one SECRET and one UNCLASSIFIED, and then proceed to copy information from the file marked SECRET into the file marked UNCLASSIFIED.   A second user S_2 with no clearance (or an UNCLASSIFIED clearance) cannot access the file marked SECRET but can access the file marked UNCLASSIFIED. Since the UNCLASSIFIED file now contains SECRET information (because of the actions of S_1), the combined actions of S_1 and S_2 violate axiom (1).

We can remove this problem by recasting axiom (1) as two axioms, one for read access and one for write access:

$$S \text{ may read from } O \text{ if and only if } CLR_S \text{ dom } CLASS_O \qquad (2a)$$

$$S \text{ may write to } O \text{ if and only if } CLR_O \text{ dom } CLASS_S \qquad (2b)$$

In this case S_1 is prevented from writing SECRET information to an UNCLASSIFIED file.

Axiom (2a) is referred to as the "simple security property" and axiom (2b) the "*-property" (pronounced star property). They form the axioms of the Bell-LaPadula Model [Bell 76] of computer security. Having determined these axioms, Bell and LaPadula then went on to develop a model of access control in the context of a typical 1970's mainframe computer system (i.e. a single mainframe computer serving a number, possibly thousands, of users via dumb terminals). This model formally proved that given a secure initial state (i.e. given that the computer system is initially configured so that axioms (2a) and (2b) are satisfied) then all transactions would maintain the validity of axioms (2a) and (2b). Bell and LaPadula then completed their work by extending their model to take account of its practical implementation in a MULTICS operating system environment.

NEW DEVELOPMENTS IN QUALITY MANAGEMENT AS A PRE-REQUISITE TO SAFETY

Denis Jackson, Data Sciences UK Limited

Abstract

This paper points out that the methods for design and production of safety-critical systems are subject to much argument, despite the considerable effort expended in their development. However there is consensus agreement on one point: that sound Quality Management (QM) is a basic pre-requisite of reliable and safe systems.

Therefore, it is argued, the role of QM in relation to Safety should be expanded. The present state of Safety regulation and QM is examined, as is the history of how this state was attained, lest we fail to profit from past lessons.

Future QM activities are proposed and discussed in the light of what will be considered in the future to be reasonable QM and what are essentially specialised Safety activities.

1   Introduction

There is a basic logic to this paper. It is:

(a)   That sound Quality Management (QM) is a basis of modern high reliability systems, whether the objective is safety or security.

(b)   There is consensus agreement on this, despite the techniques of Safety achievement, particularly software techniques, being generally deficient of consensus opinion.

(c)   Therefore we should exploit this rare consensus of opinion by identifying and developing relationships between QM and Safety.

In respect of the last point, I intend to deal with issues such as:

(a)   Where are we now in respect of Quality and the regulation of Safety?

(b)   How did we get here, as a matter of history?

(c)   Where should we go from here?

(d)   Where is the boundary between reasonable QM and more specialised Safety measures?

2   Consensus of Agreement

We live in an age of increasingly complex systems which are often only made feasible, flexible and economical through the use of software. Often the software is of an innovative nature and software is recognised as generally of poor and unpredictable reliability. Ways of improving systems reliability, particularly software reliability, are contentious and whenever a new software or safety standard is launched, there are hundreds of voices raised in dissent. In the case of the UK Ministry of Defence Draft Interim Defence Standards 00-55 and 00-56, there were approximately 1370 different comments when they were issued for public comment. Despite such general lack of consensus agreement, the outstanding point of agreement is that:

> Sound Quality Management is a basic pre-requisite of reliable and safe systems.

The reason for this appeal of QM is probably the overall discipline which it brings to the design and production processes, the applicability of its philosophies to modern high-technology systems, and its extensibility to more rigorous measures. However I do not feel obliged to justify my basic statement here because the evidence for such an assertion is overwhelming and has been stated consistently for many years in documents such as:

Cabinet Office ACARD Report [ACARD 86]
IEC SC65A, Working Group 9 Draft Standard [IEC 91], Sub-section 16.2
IEC SC65A, Working Group 10 Draft Standard [IEC 92], Sub-clause 5.1
"DRIVE" Project Report "Towards a European Standard ......"
    [DRIVE 92], Part C, Sect 5
Railway Industry Association Tech Spec 23 [RIA 91], Sub-section 4.2
Joint MOD/Industry Computing Policy [MOD 92], Para 3.8.1
Interim Defence Standard 00-31 [MOD 87a],
    RTCA DO-178A [RTCA 85], Section 7
    Defence Standard 00-16 [MOD 84]
Interim Defence Standard 00-55 [MOD 91a], Clause 11
Interim Defence Standard 00-56 [MOD 91b], Clause 7.4
Defence Standards 00-40/00-41 [MOD 87b], [MOD 89b] and [MOD 89b],
    Implicit rather than explicit
Naval Engineering Standard (NES) 620 [MOD 91c], Paras 0809
    and 0810
DCAD Technical Publication 1/77 [DCAD 79] Supporting Document No
    7, Para 7.2.

3   Where are we now?

3.1   The Regulation of Safety

We live in a time of increasing regulation by International agencies, Government departments and agencies, Local governments, Trade associations, even voluntary organisations etc. The regulatory body for a particular activity or product is not always clear and when it is identifiable it may not admit any responsibility for safety, which does require special skills. For example, the safety regulatory body needs to understand safety standards, hazard analysis, safety auditing, testing, formal proving etc, or at least to be able to conduct an orchestra of such experts. It needs to be able to detect flaws in a safety case and rectify such deficiencies until it takes an important decision, a decision which may be momentous in view of the potential consequences of accidents to modern systems. Nevertheless regulatory bodies responsible for safety do not like to be tied down to standards, preferring to retain some discretionary powers for flexibility in their operations. However good standards should be <u>demanded</u> by their users, not just thrust upon them, and there is some doubt concerning the real demand for all of the safety-related standards referenced above.

3.2 Quality Management

In recent years, suppliers have changed their attitudes from one of suspicion that QM was an overhead with an element of "gilding the lily", to recognition that QM provides a disciplined workplace, less likelihood of failed projects, less expensive redesign and rectification, and a competitive edge which appeals to purchasers.

Independent third party assessment has become the norm, avoiding the excesses of multiple assessments and providing periodic surveillance of QM.

Suppliers' organisations have thus become accustomed to being audited and have accepted the routine with openness and in the true spirit.

Purchasers' organisations have developed less readily: too often Quality is seen as entirely a supplier's responsibility and there is reluctance by purchasers to take a share in QM.

3.3 Standards

Standards are often produced concurrently by different organisations working in comparative isolation and from an initiative concerned at the apparent lack of a standard. The resulting standard may be speculative, assumed to be a "good thing to have", but unless it is demanded by users it can become a "white elephant".

One problem is that technological standards are seldom true standards but are often codes of practice. The reasons for this are, in my opinion, that:

(a) Technologies are changing so rapidly that drafting committees can never afford to stand back, survey a range of similar existing technologies, and select an appropriate technology for each application. In the early days of standardisation on tangible items such as screw threads, plugs and sockets, that process was possible, but current "standards" tend to look into the future.

(b) It is acknowledged that with modern systems, retrospective conformance testing to a standard courts failure and that it is more important to set standards for the design and production process. That tends to lead drafting committees into codes of practice covering numerous alternative techniques, none of outstanding superiority.

(c) High technology standards would receive only limited use because their rigidity would make them unaffordable in many specific applications, whereas codes of practice leave some scope for interpretation.

(d) Drafting committees tend to collate pet points which are individually sound but do not always make a cohesive or self consistent whole: greater maturity can take years to achieve.

(e) Perhaps in recognition of the preceding sub-paragraph, drafting committees understandably lack the confidence to say "lets do it this way" and know that they will be right in the areas of application of the standard.

A true standard has the benefit of curbing safety practitioners who might otherwise exceed the current limits of technology in their enthusiasm and desire to advance the "state of the art".

A noteworthy example of a successful standard has been RTCA DO-178A [RTCA 85], shortly to be superseded by DO-178B.. Produced by a sub-committee of the US Radio Technical Commission for Aeronautics (currently approximately 100 members!) it has matured from its origins in 1981 and has been recognised by many of the Civil Aviation departments which are the regulatory bodies representing over 150 Member States of the International Civil Aviation Organisation (ICAO). The standard has also been the basis of the UK MOD Interim Defence Standard 00-31 [MOD 87a], which has been used for assuring the software of UK military aircraft systems.

As aviation is an international business, RTCA effectively produced an internationally accepted standard. At the same time national and purchaser-related standards were proliferating, only to cause confusion in the developing real world of multi-national collaborative projects and multi-national selling with their attendant problems of terminology, units, contracting practices, and even national cultural differences. Proliferation of standards has now been reduced, with some national standards being withdrawn in favour of international standards, purchaser standards giving way to higher standards etc, and even good international standards such as [BSI 87] are being further improved.

Many of the QM standards produced have matured over the past ten years. Maturity has meant that the standards have been simplified and have similar content. Since Quality is defined as the extent to which a product or service meets its requirement, quality standards are dominated by the means of establishing this conformance e.g.

- Having a sound and assessed management system for product development;
- The need for a Quality Plan and for its comprehension by all concerned with the product or service;
- Having sound work instructions and standards;
- Proven observance of the Quality Plan.

3.4 The Difference between Conformance and Safety

It has to be said that conformance to Quality and Safety standards is not enough, because the system could be fully conformant to a wrong or unsafe requirement. Therefore validation of the requirement is important, but I am not satisfied that adequate validation takes place despite the expenditure of much human effort. The principal reason for this is the absence of suitable tools and some resistance to those tools which do exist. When the users' "wish list" of desirable functions for a modern large system may number 10,000 functions, it is very necessary to reconcile conflicting requirements, to collate functions into logical groups, to simplify them as much as possible and above all to ensure that a system meeting those requirements will be safe in all imaginable circumstances. In practice, the user is responsible for his actions in using a system, and major users such as airlines and railways have good operating procedures designed to keep their use of the system within safe limits.

4 How did we get here?

There have been forms of Quality and Safety regulation in existence for over a century and a half, and examples of these and the changing environment are shown in the chronological list below:

1825-1835	800 km of railways regulated by 54 acts of Parliament (15 km per Act!)
1836-1837	Additional 1600 km regulated by an additional 39 Acts.
1901	British Standards Institution formed,the first of 80 standards-making bodies.
1936	Introduction of electric signalling on railways.
1937	Air Registration Board (ARB) formed.
1944	International Civil Aviation Organisation (ICAO) formed.
1950's	Increasingly complex systems, miniaturisation, and the use of semiconductors.
pre-1970	Any concepts of Quality were governed by the use of standards, component inspection, and customers' inspectors working in industry.
1973	NATO AQAP-1 standard of Quality Management, the first of many standards reflecting increasingly complex systems and the need to design for quality.
1973	MOD Defence Standard 05-21 [MOD 73] for QM.
1974	The Health and Safety Commission (HSE and NII) established.
1979	QM Standard BS 5750 [BSI 79].
1981	RTCA DO-178 Safety-Critical Software Standard for civil aviation, since superseded by [RTCA 85].
1981	NATO Software Quality Standard AQAP-13 [NATO 81], in conjunction with AQAP-1 now [NATO 84].
1983	Naval Engineering Standard for Software [MOD 83].
1986	Naval requirements for high-integrity software [MOD 86] Annex A, since amplified in [MOD 91c].

1987	QM Standard BS 5750/ISO 9000 series [BSI 87].
1991	MOD Defence Standard 05-95 [MOD 91d], based on ISO 9000.
1991	Interim Defence Standards 00-55 [MOD 91a] and 00-56 [MOD 91b], both pioneering safety standards.
1991/92	Draft IEC Standards for Safety-Critical systems and software [IEC 91], [IEC 92], intended to be generic standards for development in particular sectors.
1991	Railway Industry Association Tech. Spec 23 [RIA 91], intended to regulate the railway signalling sector and probably the first sector-specific standard to emerge under [IEC 91].
1992	NATO Quality Standard AQAP-150 [NATO 92].

The message is one of the increasing technical complexity of regulation, of proliferation of standards in the major purchasing sector of Defence, and of the relatively backward nature of other sectors.

5 Where do we go from here?

With respect to my exceptional QM friends and colleagues, QM staff have not generally enjoyed a reputation for enterprise or original thought. However every few years they seem to jump out of their rut and I suggest that an opportunity exists now for the next jump in support of the enhancement of safety measures.

The suggested activities in this new era of QM are:

- More "facilitation" in QM, rather than having QM seen as an obstacle to progress.
- Overseeing reliability growth modelling.
- Overseeing the partitioning of software.
- Metrication of the quality of software.
- Bulk checking of code.
- Progressing from Reliability auditing (conformance-related) to Safety auditing (use-related).
- Purchaser QM.
- Ethical monitoring

I have to admit that I may be on my own in advocating such activities because I cannot detect similar ideas in the deliberations of respected bodies.

5.1 Facilitation

It would be easy for QM staff to, as some do, sit back and take the line that their role is to judge and approve quality measures, procedures, plans etc and to review objective evidence of QM actions as it is presented to them. There is a tendency, particularly amongst purchasers, to invoke every standard of conceivable relevance on the basis of "better safe than sorry". However, disentangling the relevant parts and harmonising them is a tremendous waste of human resources and involves the supplier, purchaser, user and regulatory authority. The eventually agreed hotchpotch could contain the seeds of a safety hazard if the requirements are wrong or ambiguous.

Therefore facilitation could involve more positive effort initially to specify a pragmatic package of standards, readiness to acquire a professional knowledge of the standards, and use of that knowledge to give advice readily.

Another aspect of facilitation concerns the timely processing of waivers, deviations and change orders, which will almost certainly be numerous on large projects.

Some standards concerning safety-critical systems postulate roles for imaginary players, and imply procedures. These need to be thought out if they are to be implemented by QM and Safety personnel. For example, Defence Standards 00-55 and 00-56 [MOD 91a] [MOD 91b] identify at least 42 safety-related duties of a Design Authority in addition to his normal project responsibilities.

In short, facilitation implies not a passive role for QM but one involving energy, original thought and well directed effort, but without undermining the role of QM and its traditional independence.

5.2 Reliability Growth Modelling

QM authorities could become more involved in reliability growth modelling, although I have to confess to being lukewarm concerning the value of such techniques. This is because I dispute the assumption that continued testing and defect rectification is essentially beneficial: with the normal turnover of software designers, a development team can become denuded of the original designers with their intimate knowledge of the system and therefore dependent upon system documentation. There is a possibility of fundamental errors being introduced with each rectification if that system documentation is defective.

5.3 Segregation of Safety-Critical Software

Hazard analysis essentially identifies safety-critical elements of a system, and the extent of that criticality. With small software modules, it may be possible to segregate critical and benign elements with clear interfaces between them. However, purchasers and designers are seldom satisfied with simple software systems and a modern system typically implements 10,000 user requirements. I doubt the ability to segregate the two parts, and therefore to limit the part of the design requiring special treatment, or to prove that faults in the benign part cannot be propagated into the safety-critical part. Audit of the segregation could be a QM function.

5.4 Metrication of Software Quality

Although software metrics is an underdeveloped area, at least two software tools are available to measure "software quality" in terms of "integrity", "portability" and "maintainability". One may argue about the value of such crude measures, and it is difficult to normalise and compare results, but:

(a) they stem from the aggregation of the results of many specific objective tests of the code;

(b) they can stimulate further search for reasons for anomalous results, and provide the necessary information;

(c) they meet a market demand for objective assessment of code without expenditure on labour-intensive analysis.

The operation of such tools could be a QM function. Fortunately for the safety-critical community, one of the tools will analyse Ada code whilst the other operates on 'C' (if anyone insists on writing safety-critical code in 'C'!).

5.5 Bulk Checking of Code

The same tools mentioned above permit virtually unattended operation whilst complete programs are checked, and the process is therefore relatively cheap so long as the three basic metrics are sufficient. The benefits are threefold, i.e. that:

(a) software defects are detected in parallel, in bulk, rather than by the conventional software de-bugging process of sequential detection, rectification, re-compilation and retesting;

(b) conformance to a software programming standard can be checked;

(c) all software can be checked, not just a sub-set identified as safety-critical.

The first benefit permits relatively rapid improvement of the code quality, if it is needed. However, some labour-intensive tracing through the automated analysis would be necessary.

The second benefit, automated conformance testing, must appeal to Quality staff. Typically 250 tests can be applied to the code, and the tools can be tailored to different programming codes of practice and to different objectives of the testing.

The third benefit is particularly significant in relation to safety-critical software, where segregation is an economic necessity but of dubious feasibility in any but small systems. The fact that code which is deemed to be non-critical is being subjected to scrutiny is bound to provide some confidence and in an affordable manner, and makes the dividing line less critical.

Of course QM staff must be aware of their limitations and those of bulk checking tools, e.g. spurious anomalies may be identified or more detailed independent checks may highlight problems not revealed by bulk checking. In these situations, QM staff should defer to the more experienced and better equipped safety specialists.

5.6 Safety Auditing

This is probably best carried out by a Purchaser/User QM organisation. The role exploits the auditing culture within QM, whereas any user organisation responsible for safety may not have the technical knowledge of systems or the bent for systematic auditing, and may therefore consider only the operational aspects of proposed changes in system user.

Safety auditing would consider proposed changes in operational procedures and assess them routinely from a number of aspects, e.g.: does the proposed change:

(a) create new operational roles?
(b) change an existing operational role?
(c) increase the frequency of the demand for a system service, so that an overload could develop?
(d) demand new skills or types of human response?
(e) demand more frequent human interaction?
(f) threaten to become so extensive that it becomes difficult to comprehend?
(g) weaken, or strengthen the need for, monitoring of the system or its users?

Objective answers to such general questions, and others more specific to individual users, would have helped to avert some now famous accidents.

5.7 Purchaser QM

Quality results from a two-way interaction between a supplier and a purchaser, but all too often it is seen to be the supplier's responsibility. The evidence for that is that:

(a) quantity standards are full of statements that "the supplier shall ......";
(b) a fundamental difference between an MoD standard and equivalent British or International standards is a clause giving the MoD a right of access for its "Quality Assurance Representative" so that there can be some interaction.

Some of the major purchasers such as MoD do have QM staff, but all purchasers need familiarity with QM if they are to interact with and benefit from the sophisticated QM systems which are to be found in supplier organisations. Particularly where safety-critical systems are concerned, the purchaser's QM role should not be one of checking low-level conformance to procedures but should be elevated to concern at whether the fundamental objectives of QM are being achieved.

There may even be a need for regulatory authorities to have more understanding of QM as their role becomes increasingly more complex and technically demanding.

5.8 Ethical Monitoring

The development and certification for use of a system essentially involves a compromise agreement between:

(a)   the Regulatory Authority;
(b)   the user of the system;
(c)   the purchaser of the system (if different to (b) above);
(d)   the system supplier.

A central issue is the cost of the system and its operation, and the need to justify every cost element. Thus are created the beginnings of what I call an "ethical spiral", in which the system supplier identifies a major cost element to his purchaser, who confers with the potential user and they jointly decide that the feature in question cannot be justified. When the Regulatory Authority is consulted, the ethical spiral is already well developed and the Authority has the choice of sticking to its principles, possibly laid down years ago and for a different generation of systems, or conceding to reasonable economic arguments.

The ethics of such situations are seldom clear-cut or even defined at all. UK Chartered Engineers will shortly have a Code of Practice to follow (produced by the Engineering Council) and to which the existing By Laws and Codes of Conduct of the traditional engineering institutions will be expected to be adapted. However, many developers of safety-critical systems owe no allegiance to such institutions, and software designers are notoriously loath to conform to such disciplines.

In supplier organisations it is often not clear who is responsible for professional ethics: the technical director is the most likely candidate, but I would suggest that his QM department is most likely to know of appropriate standards and codes of practice, understand their implications in the context of the supplier's business, and have the independence of action if they believe that professional ethics have been eroded unacceptably.

6   Where should Quality Management stop?

6.1   The looming issue concerns where should reasonable QM measures stop and specialised Safety measures take over. This also involves timescales, since the "special processes" of today become the norm of tomorrow. To answer this question I have tried to judge when the measures which I have proposed could naturally evolve into the norms of tomorrow. More rapid evolution might be possible under pressures from Governments; major purchasers; public opinion; or EC or National legislation.

6.2   My forecasts of implementation times, assuming acceptance of each point in principle and allowing for establishment of the organisational culture and any necessary training, are:

Facilitation	Immediate
Reliability Growth Modelling	2 years
Segregation of Safety-Critical Software	2 years
Metrication of Software Quality	1 year
Bulk checking of code	1 year
Safety Auditing	3 years
Purchaser QM	2-3 years
Ethical Monitoring	2 years

6.3 As regards what lies beyond the horizon of reasonable QM, I have considered prototyping, reverse engineering, static analysis, dynamic analysis, complexity measurement, test coverage analysis etc. Whilst they are all activities in which the independent views of QM staff would be beneficial, I thought that they were best left in the hands of designers because:

(a) the necessary tools are generally expensive and therefore less likely to be procured for the exclusive use of QM staff;

(b) if the tools are already being used by design staff, the latter should be able to react to the findings without inputs from QM;

(c) the implementation of changes is in the hands of designers, and non-essential intermediaries in the chain of discovery and rectification only tend to impede the achievement of quality and safety.

7 Conclusion

I hope that I have provoked some thoughts concerning a way ahead in the development of safety-critical systems using QM to complement specialised safety techniques.

In particular, I hope that some of the QM measures advocated will alleviate anticipated problems in:

(a) interpreting the new generation or rash of safety-critical system standards despite their immaturity

(b) filling gaps in those standards, or at least identifying the gaps, using QM expertise.

8 Acknowledgements

I thank the Directors of Data Sciences UK Limited for their support in the production of this paper, although the opinions expressed are entirely my own.

I particularly acknowledge the reviews and constructive comments of my colleagues Tom Thorne and Chris Leather.

References

[ACARD 86]   ACARD (Cabinet Office: Advisory Council for Applied Research and Development) "Software: A vital key to UK competitiveness" ISBN 0 11 630829 X, HMSO 1986.

[BSI 79]   British Standards Institution "Quality Systems - Parts 1, 2 and 3", BSI, 1979.

[BSI 87]   British Standards Institution "Quality Systems - Parts 0, 1, 2 and 3" (Identical to ISO 9000-9003), BSI, 1987.

[DRIVE 92]   DRIVE Project V1051 "Towards a European Standard: The Development of Safe Road Transport Information Systems" Draft 2, March 1992.

[IEC 91]   International Electrotechnical Commission, IEC 65A WG9 'Software for Computers in the Application of Industrial Safety-Related Systems"
Drafts only - cannot yet be referenced.

[IEC 92]   International Electrotechnical Commission, IEC 65A WG 10 "Functional Safety of Electrical/ Electronic/Programmable Electronic Systems : Generic Aspects"
Drafts only - cannot yet be referenced.

[MOD 73]   UK Ministry of Defence "Quality Control System Requirements for Industry"
Defence Standard 05-21, Issue 1, 1st January 1973 (now cancelled).

[MOD 83]   UK Ministry of Defence, Sea Systems Controllerate, "Requirements for Software for use with Digital Processors"
Naval Engineering Standard 620, Issue 1, July 1983.
(Now superseded by [MOD 91c]).

[MOD 84]   UK Ministry of Defence "Guide to the Achievement of Quality in Software".
Defence Standard 00-16, Issue 1, 9th February 1984.

[MOD 86]   UK Ministry of Defence, Sea Systems Controllerate, "Requirements for Software for use with Digital Processors"
Naval Engineering Standard 620, Issue 3, October 1986.
(Now superseded by [MOD 91c]).

[MOD 87a]   UK Ministry of Defence "The Development of Safety Critical Software for Airborne Systems"
Interim Defence Standard 00-31, Issue 1, 3rd July 1987.

[MOD 87b]   UK Ministry of Defence "Reliability and Maintainability"
Defence Standard 00-40
Part 1: Management Responsibilities and requirements for programmes and plans, Issue 2, 24th July 1987.

[MOD 89a]   UK Ministry of Defence "MOD Practices and Procedures for Reliability and Maintainability"
Defence Standard 00-41.
Part 1: Reliability Design Philosophy, Issue 2, 30th May 1989.

[MOD 89b]   UK Ministry of Defence "MOD Practices and Procedures for Reliability and Maintainability"
Defence Standard 00-41
Part 4: Reliability Engineering, Issue 2, 29th September 1989.

268

[MOD 91a]   UK Ministry of Defence "The Procurement of Safety Critical Software in Defence Equipment" Interim Defence Standard 00-55, Parts 1 and 2, 5th April 1991.

[MOD 91b]   UK Ministry of Defence "Hazard Analysis and Safety Classification of the Computer and Programmable Electronic System Elements of Defence Equipment Defence Standard 00-56, Issue 1, 5th April 1991.

[MOD 91c]   UK Ministry of Defence, Sea Systems Controllerate "Requirements for Software for use with Digital Processors.
Naval Engineering Standard 620, Issue 4, June 1991.

[MOD 91d]   UK Ministry of Defence "Quality System Requirements for the Development, Supply and Maintenance of Software"
Interim Defence Standard 05-95, Issue 1, 1st September 1991.

[MOD 92]   UK Ministry of Defence "Joint MOD/Industry Computing Policy for Military Operational Systems" Second Revision (Draft), 13th May 1992.

[NATO 81]   NATO International Staff - Defence Support Division "NATO Software Quality Control System Requirements" AQAP-13, August 1981.

[NATO 84]   NATO International Staff - Defence Support Division "NATO Requirements for an Industrial Quality Control System"
AQAP-1, Edition No 3, May 1984.

[NATO 92]   NATO International Staff - Defence Support Division "Requirements for Quality Management of Software Development", AQAP-150, 1992.

[RIA 91]   Railway Industry Association, "Safety-related software for railway signalling" (Consultative Document).
BRB/LU Ltd/RIA Technical Specification No 23:91.

[RTCA 85]   Radio Technical Commission for Aeronautics, "Software Considerations in Airborne Systems and Equipment Certification" RTCA/DO-178A, 22nd March 1985.

An Industrial Approach to Integrity Level Determination and Safety Interlock System Implementation

Victor J. Maggioli
Feltronics, Corporation
Newark, Delaware, United States

William H. Johnson, Jr.
Du Pont Company
Deepwater, New Jersey, United States

Abstract

This paper provides an overview of an industrial users methodology for determining the integrity level required for interlock applications. The method has been in operation for five years but has been continuously improved by input from, and participation in, various national (ref. 1 and 2) and international (ref. 3 and 4) standard bodies and organizations .

It is intended to provide a practical and consistent methodology for engineers to apply in the classification and implementation of safety interlock systems,

Event classification is carried out by a team of employees, with diverse experience and job assignments, during a Process Hazards Analysis (PHA). The implementation phase is prescriptive since designs exist for each integrity level that have been certified by plant experience with the scheme and the hardware. Diversity in the SISs is obtained by requiring different technologies for high integrity level solutions.

1. Introduction

This paper integrates interlock and design practices into the project safety management system so that design groups can implement cost effective, consistent, and safe interlocking schemes. The objective is to ensure that plants are safe and meet or exceed the safety requirements of the U.S.A-Occupational Safety and Health Administration.

Guidance is provided in the classification of hazardous events and the implementation of interlocks relating to personnel safety, environmental protection, and significant financial loss. It gives instructions and examples and encourages the use of the most effective interlock schemes.

The paper is divided into the following sections:

- 1) Abstract
- 2) Introduction
- 3) Definitions
- 4) Event Classifications
- 5) Classification Procedure
- 6) Integrity Selection
- 7) Implementation

- 8) Conclusion
- 9) References

2. Definitions

- Approved - Approved for the purpose/application by the project team in conjunction with process hazards specialists and/or other appropriate consultants. Project team to include design personnel as well as liaison representatives of the site on which the facility under consideration is located,

- Approved Independent Backup (AIB) - A secondary non-electrical/instrument safety layer, such as a relief valve used to prevent over pressure in a vessel. This feature will protect against a hazard if the Basic Process Control System fails. An event classification based on the existence of an Approved Independent Backup assumes that the process will be operational only when the Approved independent Backup features are operational. The backup should have the following features:

  Uniqueness - Designed to prevent/mitigate specific hazardous events. Does not have other functions in the normal operation of the facility.

  Independence Independent of other safety features such as Safety Interlocks.

  Dependability Performs with a high degree of reliability; examples, dike, relief valve, containment vessel, barricades, restricted access. Does not require human action.

  Auditability - Designed to facilitate regular validation/testing.

  Approved - Approved for each specific application by the project team.

Basic Process Control System (BPCS) - A collection of measurements, control and sequential functions, final control elements, alarms, and interlocks that normally maintain the process operation within acceptable operating limits. Interlocks that are designed to mitigate Class D & E events are part of the BPCS.

- Class A Event - A hazardous event that has the potential for serious injury to personnel and/or significant environmental impact, and has a relatively high severity and/or high frequency of occurrence, and/or is not adequately mitigated by an AIB in the judgment of the project PHR team.

271

- Class B Event - Same type of event as described for Class A above, except that in the judgment of the project PHR team#, the event has moderate severity and moderate frequency of occurrence, and/or is adequately mitigated by an AIB so as to require less interlock mitigation that Class A.

- Class C Event - A hazardous event that results in significant damages to equipment and/or significant loss of production, and has a relatively high severity, or high frequency of occurrence, and/or is not adequately mitigated by an AIB in the judgment of the project PH team, the event has low severity and low frequency, or is adequately mitigated by an AIB.

- Class D Event - same type of event as described for Class C above, except that in the judgment of the PH team, the event has low severity and low frequency, or is adequately mitigated by an AIB.

- Class E Event - A hazardous event that is determined to result in low risk to personnel, the environment, process equipment, and process operations in the judgment of the project PH team.

- Diverse - Having a different principle of operation, produced by different manufacturers, or different levels of technology produced by the same manufacturer. Different revisions of software/firmware used in a given PES are not considered diverse.

- Highway Communications - Communications between controllers within a vendor-specific architecture (ex. Honeywell - Data Highway, LCN/UCN; Modicon-Modbus II; Allen Bradley - Data Highway II). Remote input/output communications are not considered highway communications in the paper.

- Hazard- Inherent property or energy source with the potential to cause harm.

- Interlock - A system or function that detects an out-of-limits (abnormal) condition or improper sequence and either halts further action or starts corrective action. It consists of a sensing function, a control function, and a final control element. Generally part of the BPCS and is carried out with the same equipment and the same level of security as normal sequencing and other start/stop functions in the BPCS.

  Synonymous with such terms as "process interlock" and "non-safety" interlock. Failure of an "interlock" will not result in serious injury to personnel or significant environmental impact. Bee Safety Interlock.

272

Note: Interlocks operate automatically; no operator action is involved.

- Process Hazards Analysis Team - Group of individuals that bring appropriate expertise to the activity of recognizing/identifying hazards and the design features used to eliminate or mitigate them.

- Programmable Electronic Controller (PEC) - A microprocessor-based devicethat makes up the portion of a Programmable Electronic System that executes the control strategy and interlock logic. The controller receives inputs from sensors and man-machine interfaces and generates outputs to final elements. PEC is synonymous with "Distributed Control System Controller" (DCSC), "Programmable Logic Controller" (PLC), and single loop, digital controllers with integral man-machine interface.

- Programmable Electronic System (PES) - A microprocessor based system consisting of Programmable Electronic Controllers,, Sensors,, Final Control Elements, man-machine interfaces, historians, printers, etc., connected by a data highway.

- Risk - The combination of the severity of the consequence of the event and the expected frequency of the occurrence of the event.

- Safety Equipment - Approved equipment suitable to mitigate Class A and B events; safety equipment used to implement Safety Interlocks.

- Safety Interlock - A system (SIS) or function that detects an out-of-limits (abnormal) condition, or improper sequence and brings it to a safe condition. consists of a sensing function, a control function, and a final control element. The control function must be separate from the BPCS. A Safety Interlock deals with Class A & B events only, Note: Safety interlocks operate automatically; no operator action is involved.

- Safety Layers - An approved system or function specific to a hazardous event that prevents an event from occurring or protects against the consequences of an event. Examples: AIB'S, BPCS Interlocks, procedures, relief valve discharge collection systems, etc.

3. Event Classification

Process safety must be of paramount consideration in the design of any facility. There are many routes to a safe installation such as the use of control and interlock systems as well as other technology to achieve the necessary level of safety. The most effective route is through the use of inherently safe process design concepts. These utilize the concept that if the ingredients for a hazard are not present, then the process is much safer that if add-on protection is used to control the hazard.

Some industrial processing facilities, even after all efforts to employ inherently safe designs, may contain potential hazards to personnel, the environment, and the equipment used in the processing operation. By examining the hazards, with due consideration of the severity of the unmitigated consequences, the expected frequency of occurrence, and the mitigating AIB'S, the hazardous events may be classified. Based on the classification of the individual events, appropriate additional mitigation can be applied.

For the purpose of this document, events are divided into the following classes: A, B, C, D & E.

There is a need for a system to easily understand and effectively classify hazardous events. The methodology should provide solutions that meet simple as well as reasonable complex processes in a qualitative manner. For unusually complex and/or unusually sever consequences, quantitative methods such as fault tree or event tree analysis should be employed.

The severity of unmitigated consequences range from low to high and include both safety and environmental considerations. Unmitigated consequences are estimated by considering the harm that specific hazardous events can cause.

Hazardous event frequency is a key criteria to event classification. The frequency of hazardous events ranges from low to high, as follows:

The effectiveness of safety layers vary depending upon the ability of the safety layer to prevent the consequence that was identified by the PHA.

The following are examples of risk-reduction techniques:

- Reducing severity by substituting less noxious ingredients, applying scrubbers to vents, providing secondary containment, reducing batch/equipment size, reducing quantities in storage, etc.

- Reducing frequency by designing the process for more stable control, instituting procedures to reduce the frequency of or eliminate abnormal operations, reduce the frequency of rate changes, shutdowns, and start-ups, etc.

The Classification of all events involving personnel safety and environmental impacts shall be part of the **PHA**.

4.   Classification Procedure

The **PHA** team classifies each event (e.g. class **A,B,C,D** or E). The **PHA** team analyzes the non-electrical instrumentation risk reduction techniques **(AIB)** employed and adjusts the classification accordingly. This is done by one of two methods.

The first, a prescription method, by considering all severity above minor to be major, and all frequency above low to be high. The ground rules for this approval require a minimum of two separate and diverse approved independent backups (AIB,s) each capable of mitigating the risk to an acceptable level. Safety and environmental risks then remain classified Class A if no AIB's exist, or are reclassified as Class B with one AIB or re-classified as Class E with two or more AIB's. This process is precisely documented so that future process changes do not compromise the existing AIB's in a way that requires reclassification of Class E to Class A or B.

This method is used for applications with:

- process understanding
- operating experience
- prescribed design practices
- instrumentation and controls with excellent BPCS performance record, etc.

The second method uses formal Process Hazards Review (PHR) methods such as HAZOP, fault trees, event trees, etc. This approach is utilized most often for:

- new processes
- unique risk situations
- new technology
- new control approaches, etc.

5.   Integrity Selection

This process closely follows the classification procedure. There are two approaches to integrity classification. Each approach involves the use of a qualified supplier, supplier input (e.g. failure mode and effect analysis, reliability data, field performance analysis, equipment revision documentation, etc.) in both cases, technical articles and texts providing mathematical analysis of various control system configurations are utilized as comparative analysis against the methods outline herein. Rarely does the IL determined exceed the IL provided by mathematical methods.

The first approach is to utilize quantitative and qualitative IL evaluation methods in vendor and equipment selection for the BPCS. If successful, this equipment is utilized in the BPCS and its performance (e.g. operability, reliability, maintainability, etc,) is verified at this time. The results may vary between plant sites. As a result, IL evaluation is a site specific number. The data is utilized for IL evaluation for safety interlock system (SIS).

The second method involves the use of equipment with no available track record an site. This method utilizes the quantitative and qualitative methods noted above, utilizes performance data available from other sites, beta tests, performance of controls with similar technologies, etc, it will probably require a separate team analysis composed of a process hazards consultant, instrument control engineer, and supplier technical staff to understand potential common mode failure fail-to-danger, and such equipment characteristics. Understanding these problems allows development of design features to prevent the introduction of equipment induced risks on the process.

6.   Implementation

The goal of the implementation phase is to provide detailed design and installation diagrams for a few extremely effective interlock schemes. The schemes are constructed with equipment selected by the methodology outlined in the integrity section. The interlock schemes are selected in the following priority order:

- Ability to mitigate the hazardous event
- Simplicity
- Cost consideration is also given to the Life Cycle Cost of maintaining the SIS and the BPCS. A major factor in this cost is the technology supplied for control and interlocking. Experience at our sites has shown that Life Cycle Cost can be minimized by supplying plants with as few different PES technologies as possible. With this in mind, SISs acceptable for integrity levels A through E are maintained and used in application after application in a prescriptive manner. This allows a data base to be kept on actual field experience with the hardware of each control scheme.

7.   Conclusion

The design and implementation of SISs is a process that starts with event identification in a PHA. This is the most critical and time consuming step. Every effort should be made to spend the time in this step to fully cover the process under consideration using a recognized method of hazard identification. with the hazards identified, the events that can occur are listed and classified. The next step involves determining integrity level of PES's properly. This is a process that starts with the comparison of supplier information against text book data, proceeds through analysis by equipment operation, and includes site specific needs and capabilities, Since this last step is time consuming and costly, selecting a few control schemes using hardware with an excellent track record on application after application can reduce project cost and increase safety. when these factors are properly integrated, a safe and effective SIS can be designed and implemented. This IL of the SIS can only be sustained by proper audit and support.

8.   Reference

1.   *Du Pont Chemicals Interlock Design Guides*
2.   *Occupational Safety & Health Administration*   1910

3. AIChE/CCPS *"Guidelines For Safe Automation On Chemical Processes"*
4. AIChE/CCPS *"Guidelines for Chemical Process Quantitative Risk Analysis"*, New York (1989).

Unification in Uncertainty?

Bob Malcolm

Malcolm Associates Ltd

Savoy Hill, London, UK

1. Introduction

Many of the chapters in this book derive from research projects in a programme of research supported by the UK government Department of Trade and Industry and the UK Science and Engineering Research Council. (The programme is described in detail in [Malcolm 92b & 92c].)

A major ambition of this programme is to achieve understanding.

In industry, we had, on the one hand, arguments about the appropriateness of different approaches to systems development and assessment for safety-critical systems. On the other we had the emergence of new national and international standards, with implications of change or constraint on industrialists, and inhibition, at the least, on academic developments outside the main-stream.

While some standards may be timelessly independent of technology or paradigmatic assumptions, many are based on the currently prevailing system development paradigm. (Some would go further and say that most standards are necessarily obsolete!)

There is concern that new technologies such as those often labelled as 'Artificial Intelligence', and including knowledge-based systems, and neural-networks, will be either excluded from systems, because we do not have the means for their assessment, or be included, but 'under the counter', beyond even minimal controls.

Even the approaches which are fairly conventional and which have proved successful in industrial practice through the years, such as the building-block approach used in many process management systems, are criticised by some bespoke system builders on the grounds that one cannot be sure whether one is using a particular component within its correct scope of application, nor that the composition of components will have the appropriate 'emergent properties'.

(Which has recently become a topic of interest to those members of the 'formal methods' community interested in resolving this 'compositionality' problem.)

In the original workplan [Malcolm 90] a number of ideas were put forward for research to try to cut through the arguments, and to have firmer factual and theoretical bases for discussion. Could the representative power of notation play a role in the compositionality debate? Where *are* the risks in expert systems? Can we develop a sound scientific basis for the so-far intuitive concept of 'separation of concerns' - often given as an argument in support of technologies such as rule-based systems? Can the size limitations of present formal methods be overcome by embracing the notion of compositionality?

It was also suggested that to make headway with some of the problems might require the involvement of 'soft scientists', including both psychologists and philosophers - who have long discussed such things as 'representation', and 'hard scientists', such as the operations researchers, with whom systems developers may not mingle (as much as perhaps they ought?), and those on the fringes of both cybernetics and philosophy, such as those building on [Bar-Hillel and Carnap 52]. This interaction has proved difficult so far, but there are now some exciting possibilities in prospect, which may kindle the intellectual fires.

2. The problem(s)

We are interested in at least three levels of problem:

- Developing a particular system: predicting safety of the system in advance, and assessing safety before and during operation

- Assessing the weaknesses within a particular development paradigm: understanding how particular development techniques can be improved, protected against, and assessed during the development of a particular system within the paradigm

- Choosing between development paradigms, so that for a particular type of application, the most appropriate set of development techniques can be selected so as to minimise safety problems likely to be found on system developments

More generally, for the safety-critical systems community at large, we seek a generalization of this last. We would like to understand the relative weaknesses and strengths of different development techniques, so that we may better understand what is appropriate where (if anywhere!). It is this which will enable us both to cut through the arguments, and to develop more generic standards which are less likely to preclude the use of innovative technology.

3. Uncertainty - a basis for unification?

The development of a safety-critical system has been described in terms of a gradual concretion of requirements from indefinite, through realisable, to definite - that is, realised in an actual operational system [IEE 89, Malcolm 92a]. At each stage, uncertainty, in what was intended or what might be built, is removed. (Though note that there will always be *some* 'residual uncertainty' over whether what is built is what is - or should have been - intended.)

Professor Stafford Beer, in the context of decision-taking, describes a procedure for monitoring, and thereby managing, the reduction in uncertainty as decisions are taken [Beer 66]. The decision-taking which exemplifies that discussion is closest in style to the third of the 'problem levels' above - that of a 'management' choice. Unfortunately, in order to apply the approach properly, and in order to make the 'sub-choices' which the approach entails, we seem to require, not surprisingly, more information than we have at present about the properties of each technique - the second level problem above.

At the moment, the majority of the research aimed at trying better to understand this second level problem has been empirical, 'within domain', often within an organisation, and analytic. In other words, it tends to involve introspective analysis, possibly backed up by performance measures of varying objectivity and statistical validity. This is by no means intended as a criticism. It is certainly a first step to organisationally and culturally "knowing thyself". But remember Byron:

> *"How little do we know that which we are!*
> *How less what we may be!"*

To know what we may be, we must look outside our own world, to see where we fit in the greater scheme.

Now, Professor Beer's work makes use of that of the late Ross Ashby [Ashby 56], and both Beer and Ashby draw on, and elaborate on, earlier work in cybernetics [Wiener 46] and information theory [Shannon and Weaver 48].

The themes which they have in common are uncertainty (variety, in Ashby) and information, and their logarithmic measures, entropy and information (in the tighter sense of negentropy, as a logarithmic measure). Ashby was content to switch between 'raw' variety and its logarithm, as seemed most appropriate for a given circumstance. (This is a gross simplification, for brevity.)

In [Ashby 56] we find the very notion which we seek. Moreover, it is explained by means of example in the very application which we have in mind - paradigm comparison. Ashby takes a simple control application and estimates the amount of control required. He then compares between the information requirements needed to control by hand with those required to select, from a catalogue, a ready-made device which will do the job. He does not go on to compare with other options such as those with which we are regularly faced - like perhaps building such a device ourselves. So neither does he go on to look at alternative ways of building the device. In principle, though, the approach is appropriate. Each stage in our development process is a machine (in Ashby's sense) which helps us to manage uncertainty. In principle, if we could measure and characterise - even approximately - the uncertainty in different classes of problem, then we could optimise our choice of 'machines' for handling it.

Moreover, in a similar way to that in which Shannon and Weaver considered communication in the presence of noise, such an uncertainty management view could, in principle, be extended to accommodate the further uncertainty associated with the human error likely in the operation of the processes. (Where 'operation of a process' in this sense could, for example, be a software designer executing a stage in software design.)

I say "in principle" deliberately, with caution. Ashby, in order to explain the basic concepts simply but thoroughly, used 'toy' examples for the most part - amenable to mental arithmetic. Beer, on the other hand, has applied the ideas more to organizations, where variety management can often be dealt with in a broad brush manner. He uses one example of balancing the numbers of players in opposing football teams. Beer even cautions against unnecessary accuracy in the calculation of variety, such accuracy being either irrelevant or spurious. (This may not be giving a fair impression, though: he does also give a number of 'hard' examples of variety calculation.)

However, for the problems with which we are faced, we have not even identified the broad brushes which we might use. I submit that we should at least try to find them and to apply them. Can we classify, and - however crudely - quantify the uncertainties involved in different types of application, and the variety managing potential of our different techniques and wholly different development paradigms?

Furthermore, Ashby and Beer have both shown how, with respect to different systems - often simply different standpoints - the same real-world situations can embody quantitatively very different varieties. The string of symbols provided as expert knowledge to a rule-based system exhibit, symbolically, very great variety. There is much less when they are constrained to be meaningful rules, and even less when they are constrained to be rules about, say, engine performance monitoring conditions. What of the equivalent strings of design instructions submitted to the designer of a bespoke, conventionally software-engineered solution for the same task? Is this a clue to the way we might begin to see how to compare differently *structured* development paradigms?

4. Probability - a suitably broad brush?

Real-world safety cases are sometimes based on deterministic arguments that particular failures *cannot* happen. But there are deep philosophical objections to such arguments, since no-one could realistically claim perfection for *any* stage in a development chain, let alone the whole of it. Increasingly - especially since [WASH-1400] - safety cases use probabilistic arguments - or Probabilistic Risk Analysis (PRA). This is paralleled by moves to educate the public in the concept of risk - e.g. [HSE 88].

Discussions of safety, and discussions of the appropriateness of techniques either to develop safety-critical systems or to assess them, are bedevilled by confusing and apparently irreconcilable information, often of quite different types. So, in the initial workplan for the UK research programme, it was hoped that there would be research into the way that, for instance, expert judgement, test statistics, and logical analysis might be combined. It is now becoming apparent that the 'safety argument' or 'safety case' is becoming the focus of attention for this 'unification'. At least one project [Littlewood 93] is attempting to achieve this unifying reconciliation with a PRA-like approach.

Now, variety and entropy, the measures of uncertainty, are based on the likelihood of a system being in certain states, or on the range of possibilities likely to pertain as input to the systems, along with their probabilities. The notion of probability is central to all attempts to manage uncertainty. So, might the PRA be a source of unification at a higher level - a common framework for paradigm comparison? Might it even allow us to combine 'model-based' introspective characterisations of variety with empirical, statistically sound, measurement? Either way, the relationship between probabilistic assessment and varietal analysis should at least be explored.

5. Summary

From being an idealistic hope, the possibility of unification, both to reconcile different mechanisms for assessment, and, more importantly, to compare alternative paradigms, is at least a worthy vein of research.

6. Epilogue

The cybernetic references I have given here are all rather old. The reason is twofold. Firstly they are 'classics' of the discipline. Secondly, the cybernetic community, in which the ideas discussed in this paper were developed, later fragmented. Despite many commonalities, there were already some differences between this community and the General Systems movement [Bertalanffy 68]. After an initial coming together, there seems to have been a later dispersion. 'Hard' cybernetics found application in a range of 'AI'-like computer scientific topics, including neural networks, finite automata theory, and fuzzy logic. 'Soft' cybernetics became absorbed into either management science or one or other of the 'Systems' streams, some of which, such as Soft Systems Methodology [Checkland & Scholes 90] are impinging on systems development once again. General Systems Theory lives on, with a number of systems societies and journals. This is not the place for the complete listing, and certainly not for an analysis of the scene, which the participants themselves still debate. [OSG 81] contains a collection of papers representing different 'Systems' positions. Any of the current 'Systems' journals is likely to contain papers which contrast present approaches of the different schools.

One thing which I have found *quite* astonishing is the general (if not quite total) isolation of our supposed 'systems engineering' activities from all this 'systems thinking' activity of the last forty years.

References

[Ashby 56]   Ashby, W R: "Introduction to Cybernetics"
Chapman & Hall, 1956

[Beer 66]   Beer, S: "Decision and Control" Wiley, 1966

[Bertalanffy 68]   Bertalanffy, L von: "General Systems Theory" Braziller, 1968

[Checkland & Scholes 90]   Checkland P & Scholes J: "Soft Systems
Methodology in Action" Wiley, 1990

[HSE 88]   Health and Safety Executive: "Tolerability of risk from nuclear
power stations" HMSO, 1988.

[IEE 89]   Institution of Electrical Engineers and the British Computer
Society: "Software in safety-related systems: a report prepared by a joint project
team" (isbn 0852963 91 2) IEE, 1989

[Littlewood 93]   Littlewood, B: "The need for evidence from disparate sources,
to evaluate software safety" In Redmill, F & Anderson, T: "Directions in
Safety-Critical Systems", Springer-Verlag, 1993

[Malcolm 90]   Malcolm, R E: "Workplan for the JFIT Safety Critical Systems
Research Programme" Department of Trade and Industry, 1990

[Malcolm 92a]   Malcolm, R E: "Software in safety-related systems: basic
concepts and concerns" In: Bennett, P.A. (Ed.) "Safety aspects of computer
control" Heinemann-Butterworth, 1992

[Malcolm 92b]   Malcolm, R E: "The JFIT Safety Critical Systems Research
Programme: origins and intentions" In Redmill, F & Anderson, T: "Safety-Critical
Systems - Current Issues, Techniques, and Standards" Chapman & Hall, 1993

[Malcolm 92c]   Malcolm, R E: "The UK Safety Critical Systems Research
Programme" In: Proceedings of the CSR conference on Software Safety,
Luxembourg, 1992. Centre for Software Reliability, 1992

[OSG 81]   The Open Systems Group of the Open University: "Systems Behaviour" 3rd Edn. Paul Chapman, 1981

[Shannon & Weaver 49]   Shannon CE & Weaver W: "The Mathematical Theory of Communication" University of Illinois Press, 1949

[WASH-1400]   "An assessment of accident risks in US Commercial Nuclear Power Plants: Reactor Safety Study" US Nuclear Regulatory Authority, 1975

[Wiener 48]   Wiener N: "Cybernetics" Wiley, 1948

AUTHOR INDEX